Optoelectronics: Modern Developments

Optoelectronics: Modern Developments

Edited by **Rodney Lappin**

LANRYE
INTERNATIONAL

New Jersey

Published by Clanrye International,
55 Van Reypen Street,
Jersey City, NJ 07306, USA
www.clanryeinternational.com

Optoelectronics: Modern Developments
Edited by Rodney Lappin

International Standard Book Number: 978-1-63240-405-3 (Hardback)

Contents

Preface

This book discusses the advanced concepts in the field of optoelectronics. Optoelectronics focuses on the study and function of electronic tools that release, identify and control light. As a field of study, it has widely flourished internationally and enabled many of the currently used conveniences. Because of this universality, novel functions and optical phenomena have led to further modernization. This book covers current achievements by experts around the world. A lot of developed and developing nations have contributed to this attempt. Chapters in this book are written by prominent scientists working in USA, Japan, India, as well as through the joint participation of the US and Moldova scientists. This book also describes the methods of production of optoelectronic devices. It also includes an analysis of operation and areas of application of such devices.

Significant researches are present in this book. Intensive efforts have been employed by authors to make this book an outstanding discourse. This book contains the enlightening chapters which have been written on the basis of significant researches done by the experts.

Finally, I would also like to thank all the members involved in this book for being a team and meeting all the deadlines for the submission of their respective works. I would also like to thank my friends and family for being supportive in my efforts.

Editor

Aromatic Derivatives Based Materials for Optoelectronic Applications

Florin Stanculescu and Anca Stanculescu

Additional information is available at the end of the chapter

1. Introduction

In the last decades a high interest has been paid to the field of organic materials for electronic and optoelectronic devices as potential candidates for replacing the more expensive, energy consumer and polluting technologies involved by inorganic semiconductor devices.

One of the most important advantages of the organic materials is the possibility to modify and optimize their molecular structure using the advantages of the design at the molecular level and the versatility of the synthetic chemistry with the purpose to tune their properties and make them adequate for a well defined optic, electronic or optoelectronic application.

Organic light emitting diodes (OLED) are interesting for applications in the full-colour flat panel displays and new generation of lighting source as an alternative to incandescent bulbs and compact fluorescent lamps. The OLEDs based technology has a large area of applications from small mobile phone displays to TV and monitors because they show two important advantages compared to the competitive liquid crystals based technology: high brightness and wide viewing angle.

Since the discovery of luminescence in anthracene [1], the crucial moment in the development of the OLEDs technology was the realisation of the first organic bilayer structure able to emit light at low applied voltages [2]. After that, different organic multilayer structures have been tested to improve carrier injection, carrier transport and radiative recombination with the purpose to increase the OLED efficiency and lifetime [3-5]. The performances of the devices can be enhanced either by the selection of an adequate architecture, such as multilayer structure or by doping, by controlled impurification of the organics.

The research was focalised on different topics such as: effect of doping of the organic semi-conductor to increase the "transparency" of the energetic barrier to the injection of electron from the contact [6], influence of the trapped and interfacial charges generated in multilayer organic heterostructures on the properties of the device [7], charge tunnelling in multilayer stack and at the interface between organic and anode [8], influence of the thickness and doping of the emission layer on the properties of OLEDs [9], injection of the charge carriers from the electrodes and their migration in correlation with different types of cathodes [10-13], transport phenomena in organics [14; 15], stacked geometry for efficient double-sided emit-ting OLED [16], graded mixed layer as active layer to replace heterojunction in OLEDs [17]. The awarding in 2000 of the Nobel prize for researches in the field of conducting polymers has stimulated the development of OLEDs based on polymeric materials, application em-phasised years before [18], opening the way of reducing the applied voltages (< 10 V) and increasing the brightness and lifetime.

The development of the technologies in the field of the structures for optoelectronic applica-tions based on organic compounds is dependent of the development of fundamental and ap-plied knowledge of all the optical and electrical processes involved, because many particularities of the organic solid state are not yet well known and understand. This is a re-al challenge because the number of the organic luminescent compounds is much larger than the number of inorganic compounds and it is continuously increasing. To increase the quan-tum efficiency, lifetime and thermal stability of these devices requires the separate optimisa-tion of the generation, injection and transport of the charge carriers and their controlled recombination in different layers. The electrical properties of the OLED are controlled by the mobility of the charge carriers and the heights of the barriers [19], whereas the optical prop-erties by the refractive index mismatches at the glass/air and organic/ITO interfaces that generate the trapping of a large fraction of the light by the mechanism of total internal re-flection into glass and ITO [20].

Therefore the organic materials must be designed and selected in such a way to show spe-cial properties to satisfy these requirements. Organic luminescent materials can be divided considering their molecular structure and the macro scale organization. From the first point of view there are low-molecular compounds characterised by the possibility of high purifi-cation, easy vacuum deposition, high quantum yield fluorescence and large variety and high-molecular compounds (oligomers and polymers) characterised by mechanical strength, flexibility and luminescence over various spectral regions from near UV to near IR but, by small quantum yield of fluorescence. From the second point of view, macro scale organiza-tion, there are bulk crystalline organic and organic thin films and heterostructures to be used in devices' fabrication.

A special attention will be paid in this chapter to investigate the properties of bulk and thin films organic compounds showing both good optical, including luminescent, and transport properties for potential optoelectronic applications.

2. Bulk aromatic derivatives for optoelectronic applications

Organic luminescent solids are attracting increasing interest in various field of application from optoelectronics to photonics. The interest in studying organic crystals is justified by the perspective to use these materials as a crystalline host matrix both for organic and inorganic guests (dopants) for developing new classes of materials combining the advantageous properties of both components host and guest. The organic matrix can assure an efficient fluorescence mechanism, can assure simple methods for processing and can contribute to electrical transport. On the other hand, the dopants could increase the charge carrier mobility and improve the emission properties and thermal stability of the organic.

Organic molecules containing electrons occupying nonlocalised molecular orbitals and strongly conjugated systems such as aromatic compounds, dyes, show important luminescence in solid state. This radiative emission involves transitions inside very well shielded systems of π-electrons. By light absorption, an electron is transferred to an antibonding π orbital on the lowest singlet excited state with a lifetime of 10^{-6}-10^{-9} s, from which it decays by fluorescence emission.

The perspective to tailor the specific physical properties of a molecular solid by guest particles (dopant) embedded in the crystalline organic matrix is very attractive, but not so accessible because some complications can appear both from the crystalline structure and/or dopant sites.

Special research has been devoted to the growth of organic crystals doped with rare earth metallic ions to prepare materials for luminescent and laser applications and benzil doped with Cd^{2+}. The properties of the host/guest systems based on organic crystals depend on the crystalline perfection and chemical defects.

Growth of large and structural good organic crystals at good ratio cost/properties is very important for theoretical understanding of the phenomena taking place in organic solid state and development of new organic-organic, organic-inorganic materials for a target application. The main limitations in large-scale using of aromatic derivatives as crystalline matrix are correlated with the requirements for crystals growth, which involves identification of particularised solutions to overpass the low melting point, supercooling and low thermal conductivity of organic compounds.

Substituted aromatic molecules are a class of organic materials containing weakly coupled, strongly polarisable delocalised π electrons. Concerning the bulk organic crystals, our interest was focalised on aromatic derivatives that contain one and two aromatic rings and substituent groups which disturb the symmetry of the π-electrons cloud, such as meta-dinitrobenzene (m-DNB)/ $C_6H_4N_2O_4$ and benzil/ C_6H_5-CO-CO-C_6H_5, characterised by large transparency domain and good fluorescence emission.

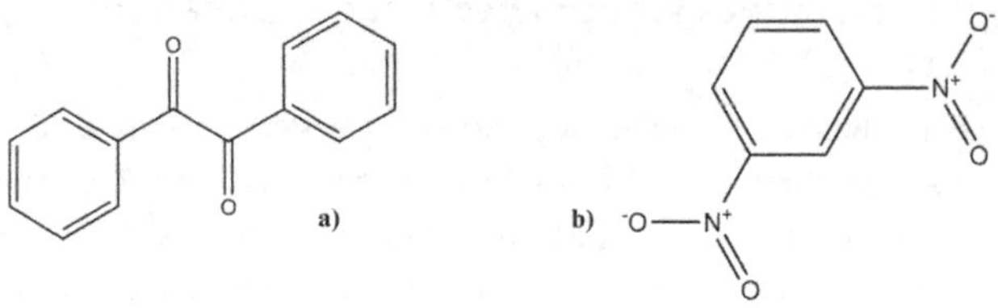

Figure 1. Benzil (a) and m-DNB (b) molecular structure [21].

Benzil with the molecular structure presented in Figure 1a is an uniaxial crystal that belongs to the space-group D_3^4 or D_3^6 and it is known as "organic quartz" being isomorphic with α quartz. By similarity of the microstructures developed in quartz through the diffusion of metal atoms could be of great interest to study benzil as matrix for composite materials and the effect of dopant atoms on the matrix properties.

Meta-dinitrobenzene with the molecular structure presented in Figure 1b is a negative biaxial crystal that belongs to the point group symmetry mm2 and space group Pbn2$_1$.

We have developed some investigations on the effect of dopant on the emission properties (shape of the spectra, position of the peaks) of the solid-state aromatic compounds by comparison with the emission properties of the pure organic matrix. We have also evidenced the differences between the influence of the inorganic dopant (iodine, sodium, silver) and/or organic dopant (m-DNB, naphthalene) on the luminescence of bulk m-DNB and benzil samples.

2.1. Aromatic derivatives crystal growth

The source materials used in crystal growth must be of high purity and the purification of organic compounds is a very long process. m-DNB was purified by three methods: chemical purification, vacuum distillation and two steps directional freezing in a horizontal configuration: length of the melted zone=2-3 cm, average travelling speed=2.5 cm/h [22; 23].

Some factors have contributed in the selection of Bridgman-Stockbarger method in vertical configuration to grow m-DNB crystals: low melting point, low vapour pressure and no decomposition at the melting temperature. The supercooling tendency of the organic compound was counteracted in a special design system with two zones (a hot zone: 110-115 °C generated in a furnace that assures the melting of the charge and a cold zone: 50 °C generated by a thermostat) characterised by a steeper thermal gradient at the growth interface created by an oil bath. It is very important to correctly positioning the growth interface compared to the interface air/ oil. To allow the dissipation of the high melting heat in the organic material characterised by low thermal conductivity, the ampoule containing the crucible sealed under vacuum is moved slowly in the thermal field. The Teflon dismantle or undismantle crucible containing the organic compound powder has a special bottom configuration (a capillary tube with a diameter of 1 mm) to generate the nucleation and to favour the selection of the growth direction. Details about the experimental configuration are given in Figure 2.

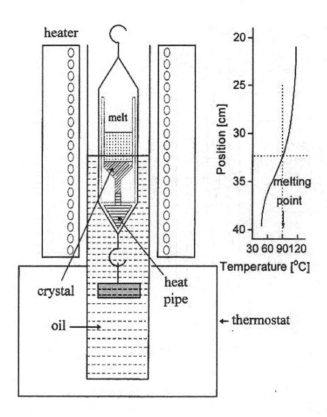

Figure 2. Experimental set-up for the growth of m-DNB crystals and the corresponding thermal profile [23].

The two parameters, thermal gradient at the melt-crystal interface of 4.5-5 °C/cm; 8.5-9 °C/cm and average moving speed of the ampoule in the furnace of 1-1.5 mm/h; 1 mm/h are very important for the crystal growth process because are determining the shape of the solid-liquid interface and position of the growth interface with effect on the properties of the crystals.

A similar configuration, presented in Figure 3 has been used for the growth of the benzil crystals. Some differences result from the fact that benzil is characterised by a weaker adhesion to the quartz wall than m-DNB (the use of a Teflon crucible being not necessary in this case) and from the necessity to assure the control of the nucleation and solidification processes in a configuration without crucible by the use of a conical shape of the ampoule tip with a narrower zone [24].

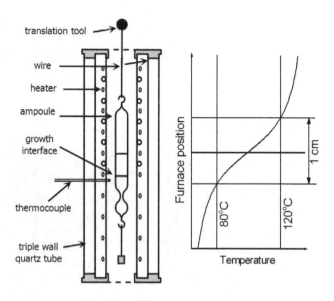

Figure 3. Experimental set-up for the growth of benzil crystals and the corresponding thermal profile [24].

A very important parameter in the process of crystals growth is the temperature, which has two counteracting actions:

1. low thermal gradients at the growth interface are necessary to prevent the generation of mechanical defects, favoured by the accumulation of tensions inside the crystal, like cracks;

2. steep gradients are necessary at the same growth interface to counteract the supercooling effect and the tendency to a facetted growth morphology [24; 25].

In general, the organic compounds are characterised by a low thermal conductivity in solid phase and high values of the solidification enthalpies that must be liberated during the crystallization process. In the system matrix/solvent can be developed many flow cells leading to non-uniform distribution of the dopant in the matrix. Constitutional supercooling characterises the doped organic melt because the freezing front rejects the particles of dopant, which can accumulate in front of the moving solid-liquid interface, the equilibrium freezing temperature of the adjacent liquid is above the actual temperature and the gradient of the equilibrium temperature is:

$$\Delta T_e = \Delta C \cdot m \qquad (1)$$

where ΔC=concentration gradient at interface; m=slope of the liquidus curve. $\Delta Cm > 0$ by convention, and for molecules rejected at the interface, that decrease the melting temperature, m<0 [26].

The problem of the growth interface stability is very important because the growth interface has effect on the quality of the obtained crystals. Our benzil/dopant system was analysed using the Mullin-Sekerka criterion [26-29], that fixes the limits of the stable growth and the conditions necessary to initiate instabilities in the growth system, and is defined by the following relation:

$$\frac{\left(V \cdot \rho_m \cdot \Delta_f H - \left(k_m - k_s\right) - \Delta T\right)}{\left(k_m + k_s\right) \cdot \left(\Delta T + \Delta C \cdot m\right)} \geq 1 \tag{2}$$

where V=ampoule moving speed; ρ_n=melted benzil density; k_m=melted benzil thermal conductivity; k_s=thermal conductivity of benzil crystal; $\Delta_f H$=solidification enthalpy; ΔT=thermal gradient; ΔC=concentration gradient at the growth interface; m=slope of the liquidus curve. For the growth of pure benzil crystals the stability condition became:

$$\Delta C \cdot m \leq 10.0727 \cdot V - 0.4382 \cdot \Delta T \tag{3}$$

with $\Delta T < 0$ ($T_{final} < T_{initial}$ in the solidification process).

In benzil/dopant system for the given experimental conditions, the stable and unstable growth zones were delimited by the curves $\Delta C \cdot m = f(|\Delta T|)$ when V=constant or by the curves $\Delta C \cdot m = f(V)$ for $|\Delta T|$=constant. In the first case, as can be seen in Figure 4, at high concentration gradients (ΔC) the system moves through the unstable growth zone situated above the curve given by equation (3). For a given thermal gradient at the growth interface small variations in the interface moving speed have no significant influence on the area of the stable growth zone. The main consequences refer to an increase in the morphological instabilities and in crystal's homogeneity. In the second case presented in Figure 5, the area of the stable growth zone increases with the increase of the thermal gradient for a given moving speed of the growth interface, the system remaining in the stable growth zone even for high concentration gradients at the interface.

For benzil crystals $k_m > k_s$ and as consequence the interface is more stable because the term $\left(k_m - k_s\right) \cdot \Delta T \Big/ \left(k_m + k_s\right)$ in equation (2) assures a large range of values situated in the stable growth zone.

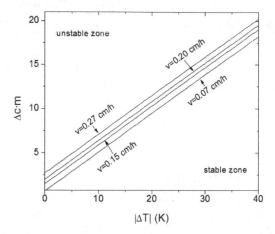

Figure 4. Stable (bellow the curves) and unstable (above the curves) growth zones for the system benzil/dopant in Bridgman-Stockbarger configuration delimited by the curves:$\Delta C \cdot m = f(|\Delta T|)$, V=constant [24].

These considerations are very important in choosing the parameters for a stable growth generating homogeneous crystals. All the studied systems based on pure and doped benzil and m-DNB matrices are similar from the point of view of the solidus-liquidus interface stability criterion because the ratios $|\Delta T| / V$ are comparable [26; 24].

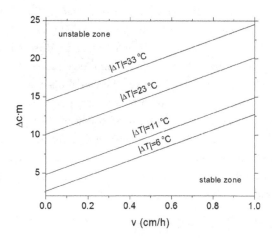

Figure 5. Stable (bellow the curves) and unstable (above the curves) growth zones for the system benzil/dopant in Bridgman-Stockbarger configuration delimited by the curves:$\Delta C \cdot m = f(V)$, $|\Delta T| = constant$ [24].

The same parameters, thermal field and the interface moving speed are important in the engulfment or rejection of the dopant particles in the crystallisation front. Compositional variations and growth micrononhomogeneities (named striations) appear because of the layer situated in front of the interface, which is enriched in foreign particles by rejection. The factor which influences the incorporation of the dopant atoms/molecules in the matrix are: the shape, volume and intermolecular bonds of the dopants' molecules. A free space around 2.9 Å has been evaluated considering the molecular structure and the geometry for both benzil and m-DNB [24]. The diameter of the dopant atoms favours the incorporation in interstitial positions and the incorporation is facilitated by the local deformation of the organic lattice characterised by weak Van der Waals forces [25; 24].

Sodium atoms could be incorporated interstitially with difficulty because the atomic diameter is greater than the diameter of the free space in benzil solid state, while silver and iodine with a diameter smaller than the free space could be easily interstitially incorporated. In the case of metals characterised by a high first ionisation enthalpy is sustained the generation of clusters, which can create difficulties in atoms incorporation (like iodine or silver).

For sodium atoms the situation is a little bit more complicated because of the high reactivity of sodium [25]. The interaction between the organic molecules and the alkali and alkali-earth metal atoms is determined by a chemical reaction leading to an organometallic complex or a charge transfer generating anion-cation pairs. The first situation is characterised by a very low probability because the hydrogen in benzil has not a very strong acid character to be directly substituted by alkali metal atoms and the crystal growth in closed sealed systems under vacuum reduces the possibility of sodium oxides formation. The charge transfer in sodium doped benzil crystals is caused by a nonbonding Van der Waals force present in all the organics and a dative bonding force corresponding to the situation in which two Na donor atoms could give each the outer 3s electron to two oxygen atoms from the carbonyl acceptor group in benzil. Because the ionisation energy for most alkali metal atoms is low (around 4-6 eV) the electron transfer from Na-guest to benzil-host is allowed without the formation of a covalent bound [25].

In the case of an organic dopant the situation is completely different and depends on the matrix. Because the organic molecules are big and can accommodate in the host lattice only substitutionally and not interstitially and must respect the condition for solubility in solid phase and the criterion for the geometrical similarity between the molecule of the dopant and matrix [30; 31]. If there are geometrical differences, the substitution is less probable and the microinclusions of dopant can generate distortion of the lattice and cracks.

The geometrical similarity is measured by the overlapping factor represented by the ratio between the unoverlapped and overlapped volume of the matrix and dopant [32]. This introduces a limitation of the doping level which can be allowed by each host/guest system. The volume of m-DNB, benzil and naphthalene molecules have been estimated supposing a spherical shape of the molecule and taking into account the length of the chemical bounds. Because the calculated unoccupied volume is much greater than the occupied volume, the possibility for m-DNB to replace benzil molecule and be included substitutionally in the lattice is very small. The m-DNB molecules, which are not completely dissolved in benzil, segregate and generate microinclusions that favour the light scattering. The smaller geometrical differences between benzil and naphthalene generate weaker segregation effect.

2.2. Optical properties of bulk aromatic derivatives

The segregation effect of the dopant was investigated experimentally by UV-VIS measurements. The transmission of benzil doped with m-DNB sample is lower than the transmission of benzil doped with naphthalene sample (for the same thickness ~2 mm) as can be seen in Figure 6, suggesting a stronger segregation of m-DNB than naphthalene in benzil matrix. The segregation of iodine in benzil matrix with effect on the homogeneity, reflected in UV-VIS spectra, is stronger in the presence of another dopant (naphthalene or m-DNB) and less significant in the absence of any other organic dopant as presented in Figure 7.

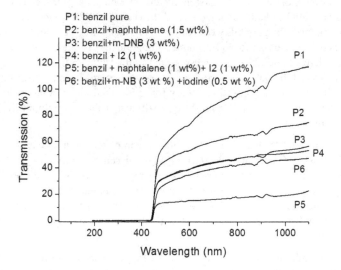

Figure 6. The effect of dopant on the UV-VIS transmission spectra of benzil matrix [24].

UV-VIS absorption spectra of pure benzil and benzil doped with Ag or Na, presented in Figure 8, Figure 9 and Figure 10, have specific shapes characterised by a narrow peaks structure at wavelength < 450 nm.

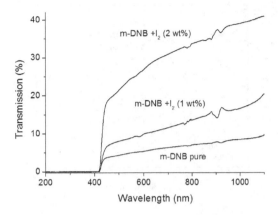

Figure 7. The effect of dopant on the UV-VIS transmission spectra of m-DNB matrix [24].

The UV-VIS spectrum of pure benzil, presented in Figure 8, preserves the pattern by doping with Ag, which is not interacting with benzil molecules. As result, the fundamental absorption edge is not affected and preserves the narrow peaks structure. The peak situated around 380 nm is correlated with some particularities of the benzil molecular configuration and is attributed to the absorption on the dicarbonyl groups [33; 21].

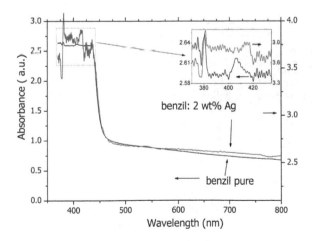

Figure 8. Comparison between the absorption spectra in bulk samples of pure benzil and benzil doped with Ag [25].

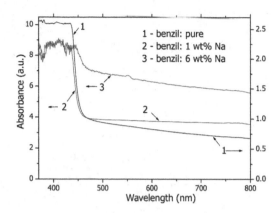

Figure 9. Comparison between the absorption spectra in bulk samples of pure benzil and benzil doped with Na [25].

At the contrary, doping with Na has introduced important changes in the shape of the fundamental absorption edge in benzil, as can be observed in Figure 9, a large structured band replacing the narrow peaks. This can be explained by the light scattering on the nonhomogeneities of the doped benzil or by changes in the forces acting between the host and guest [25].

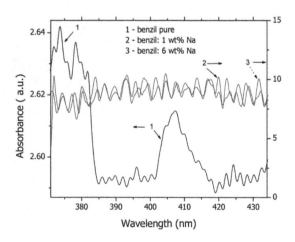

Figure 10. Absorption spectra of bulk pure and Na doped benzil matrix. Detail between 375 nm and 435 nm [25].

From the transmission data near the fundamental absorption edge processed using a linear-power model characterised by a formula obtained by superimposing a linear function and a power function [25]:

$$\alpha = a + b\left(E_g - c\right)^d + mE \tag{4}$$

or

$$\alpha = a + b\left(c - \lambda_g\right)^d + m\lambda \tag{5}$$

where: α=absorption coefficient; c=band gap energy, E_g, or edge of the fundamental absorption, λ_g, respectively, d=coefficient that depends on the light absorption mechanism and (a +mE) or (a+mλ) respectively, define all the other parasitical processes, including scattering of light on the nonhomogeneities of the sample and affecting the band to band absorption mechanism.

The light absorption process is characterised by the optical band gap, which in benzil has been evaluated at E_g=2.65 eV, emphasising the wide band gap semiconductor character of crystalline benzil. The narrowing of the optical band gap by introducing energetic levels in the band gap in pure crystal can be a consequence of physical defects or controlled doping that can generate chemical or structural defects.

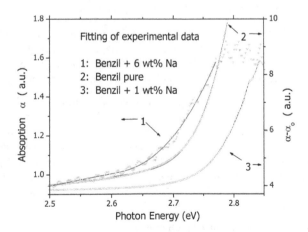

Figure 11. Fitting of the experimental data for bulk samples of pure and Na doped benzil [25].

As can be seen in Figure 11, the narrowing in the optical band gap is correlated with the presence of physical defects because the large radius metal atoms disturb the organic lattice and the effect of the dopant is hidden by the structural imperfections [25]. The impurities migrate and concentrate at these defects, such as grain boundaries, twins, dislocations.

For benzil doped with Ag presented in Figure 12, the situation is different and the optical absorption involves energetic levels from the band gap associated with the generation of cluster, as a consequence of the high first ionisation energy and weak reactivity of Ag.

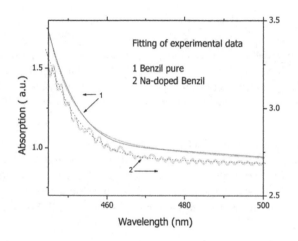

Figure 12. Fitting of the experimental data for bulk samples of pure and Ag doped benzil [25].

At excitation with λ=335 nm, the bulk sample of m-DNB shows a high, broad emission peak presented in Figure 13 [34], which correlated is with the radiative decays from the first excited energetic level, peak situated at 2.85 eV with a shoulder at 2.95 eV generated probably by the radiative decay from another vibrational level of the same excited energetic level.

A modification of the spectrum could be evidenced when m-DNB is doped with iodine, the position of the emission slightly moving through shorter wavelengths, the difference between the peak and shoulder being attenuated by the increase of the iodine concentration from 1 wt % to 2 wt %. The blue shift of the emission peak can be associated with the migration and trapping of the exciton on the defect zones characterised by slightly higher energy compared to m-DNB without defects. Despite the strong interaction between the molecules, the energy of the level associated to the defect still remains under the energy of the exciton level.

The peak situated at 2.8 eV evidenced both in pure and doped m-DNB can be obtained by a radiative decay from the lowest excited triplet state to the ground state. This forbidden transition became possible by the relaxation of the selection rule in m-DNB under the effect of the vibrational interactions. The triplet state can be reached by a radiationless "intersystem crossing" process from the excited singlet state.

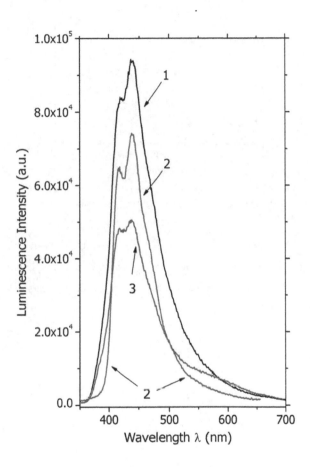

Figure 13. Luminescence spectra of m-DNB crystals: (1) pure m-DNB; (2) m-DNB doped with 1 wt % iodine; (3) m-DNB doped with 2 wt % iodine [34].

The m-DNB molecules contain oxygen atoms with lone electrons pairs which can be promoted to an unoccupied π orbital and give rise to a (n, π^*) singlet excite state and a triplet excited state, with an energy lower than the energy of usual (π, π^*) state. The (n,n*)state in m-DNB is the lowest excited singlet state that favours the radiationless processes as "intersystem crossing" or "internal conversion" to a lower excited energetic level. The (n, π^*) transition in m-DNB is confirmed by the small value of the "singlet-triplet" splitting evaluated from the correlation of absorption and emission data [34].

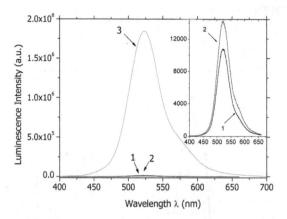

Figure 14. Luminescence spectra of benzil crystals: (1) pure benzil; (2) benzil doped with 1 wt % iodine; (3) benzil doped with 2 wt % iodine [34].

The luminescence spectra of pure benzil presented in Figure 14, shows an emission peak situated at 2.37 eV generated by the lone electron pairs of oxygen atoms in carbonyl groups emitting only from planar configuration, on which are localized the emission transition involving (n, π^*) states.

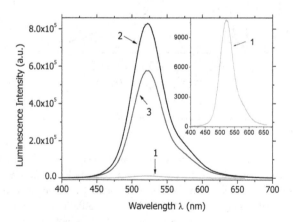

Figure 15. Luminescence spectra of benzil crystals: (1) pure benzil; (2) benzil doped with Ag (2.4 wt %); (3) benzil doped with Na (1 wt %) [34].

The doping with other metallic impurities such as silver (2.4 wt%) or Na (1 wt %) has not modified the sharp peak situated at 2.37 eV, which is present in pure benzil, as it is emphasised in Figure 15, and this peak could not be correlated with an exciton trapping mechanism [34]. The peak at 2.37 eV could be generated by the radiative decay from the excited triplet state (T_1) to the ground state (S_0), which is a transition forbidden for separated molecules becoming allowed through the vibrational interactions when the molecules are coupled in the solid state, in a crystalline lattice [23; 35; 36].

In Figure 16 is presented the spectrum of benzil doped with m-DNB. The benzene derivative, m-DNB, is active itself and has a direct action on the benzil matrix. The peak assigned to m-DNB is situated at 2.97 eV in the high-energy range of the emission spectrum. The preferential excitation of m-DNB molecule can be explained by the lower position of the excited singlet state (2.9 eV) in m-DNB compared to benzil (3.25 eV). At the contrary, naphthalene has not any significant influence on the emission spectrum of benzil presented in Figure 17, because the first singlet excited state in naphthalene is situated at ~ 3.84 eV and the triplet state at ~ 2.64 eV higher than the corresponding energetic levels in benzil (3.25 eV and 2.3 eV respectively). Therefore the peak observed in the emission spectrum of naphthalene doped benzil is obtained by the radiative deexcitation of benzil matrix.

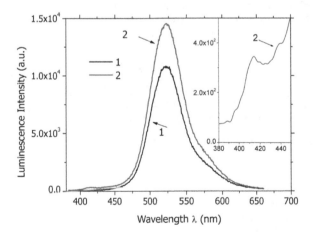

Figure 16. Luminescence spectra of benzil crystals: (1) pure benzil; (2) m-DNB doped benzil (3 wt %) [34].

As can be seen from Figure 17, the simultaneous doping with naphthalene and iodine has no significant effect on shape of the emission spectrum of benzil.

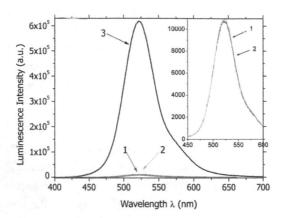

Figure 17. Luminescence spectra of benzil crystals: (1) pure benzil; (2) benzil doped with naphthalene (1.5 wt %); (3) benzil doped with naphthalene (1 wt %) and iodine (1 wt %) [34].

3. Aromatic derivatives thin films for optoelectronic applications

The major problems for large-scale application of crystalline matrices from these aromatic derivatives materials are associated with the difficulties to grow (low melting point, super-cooling, and low thermal conductivity), process (weak mechanical properties determined by weak bonding forces between molecules) and doping organic crystals to assure an homogeneous distribution of the guest atoms.

Crystalline organic films are preferred in a variety of applications because of the complexity of the processes involved and long time necessary to grow bulk organic crystals. The use of thin films is more promising because represents an optimum between the cost of manufacturing and properties of interest for special and oriented applications.

Investigation of the properties of organic thin films is a very important aim because these films are components of the organic heterostructures as fundamental elements of any organic devices. It is also necessary to investigate the properties of heterostructures because the junction between two different semiconductors (organic/organic; organic/inorganic) or between a metal and a semiconductor (metal/organic) is the key building block of any modern electronic, photovoltaic and optoelectronic devices. Heterojunction technology has known a continuous development from the first heterojunction transistor, realized by Bardeen in 1948 at Bell Laboratory [37], to p-n junction transistor, realized by Schockley in 1949 [38], and to nowadays devices based on multilayer heterostructures.

The organic compounds which have been investigated as components of organic compound based heterostructures are:

a. 3,4,9,10-perylenetetracarboxylic dianhydride (PTCDA);

b. Zinc phthalocyanine (ZnPc);

c. tris(8-hydroxyquinoline) aluminium (Alq3) and

d. 5, 10, 15,20-yetra(4-pyrydil)21H, 23H-porphyne (TPyP).

PTCDA is known as having p type conduction while Alq3 and TPyP are characterized by n type conduction. They molecular structure is given in Figure 18.

ZnPc is an electron donor forming highly ordered layer, with a broad transmission window in visible region of the spectrum [39].

In PTCDA, an electron acceptor, the interaction between the π-electrons systems is favored by the planar molecule and the perpendicular stacks of molecular planes [40], which determine a quasi-one-dimensional molecular crystal structure [41].

Alq3 shows different stereoisomers (median, facial) determined by the mutual orientation of the ligands of hydroxyquinoline, which show different symmetries and as consequence different properties [42; 43].

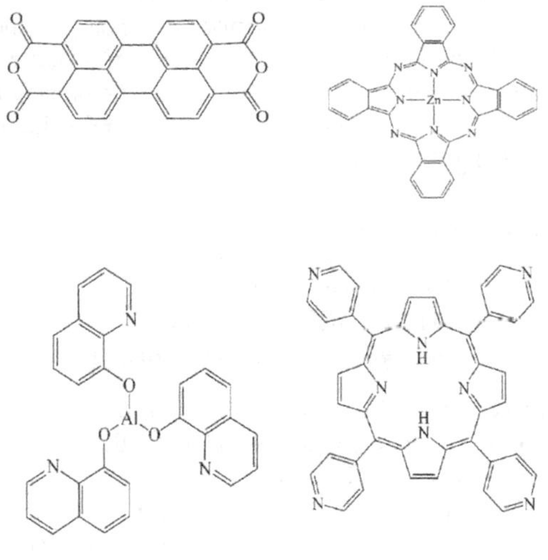

Figure 18. PTCDA (a), ZnPc (b), Alq3 (c) and TPyP (d) molecular structure.

TPyP is a non-metallic porphyrin with an increased electron affinity obtained by the substitution of phenyl group by pyridyl group determining the n type conduction. The basic structure of porphyrin consists in four pyrrolic entities linked by four unsaturated methane bridges with a skeleton showing an extended π-electrons system assuring a large spectral range for light absorption [44].

3.1. Preparation of the aromatic derivatives thin films

3.1.1. Aromatic derivatives thin films preparation by directional solidification process

It is not very easy to obtain organic thin films because of the same difficulties which affect the preparation of organic crystals and the quality of the organic layer is strongly influenced by the method, which have been selected to grow the film. For example there is a high concentration of structural defects in the benzil thin films which have been grown by a rapid directional solidification process, characterised by a non rigourous control of the thermal regime compared to the crystal growth process and these defects have caused the red shift of the emission peak. It is very difficult to grow, by vacuum evaporation, thin films of organic compounds characterized by a melting point $T_m<100$ °C (including benzil and m-DNB) because the heating of the substrate during the evaporation process can favor a strong desorption of the organic molecules from the substrate. By the directional solidification process could be prepared organic thin films between two substrates, like quartz or glass substrates, during a rapid thermal solidification characterized by a temperature gradient for solidification, $\Delta T>50$ °C, necessary to counteract the supercooling phenomenon [45]. Crystalline fragments from organic ingots of pure and doped m-DNB ($T_m=89.9$ °C) or benzil ($T_m=95$ °C), grown by Bridgman-Stockbarger method presented in paragraph 2.1, have been melted between the substrates by the hot plate technique and after that rapidly frozen by the cold plate technique obtaining films with a columnar structure with large dendrites branches in the plane of the film determined by the low thermal conductivity and anisotropy of these organic compounds. The thickness of the films has been evaluated (using the density in solid state) from geometrical considerations presuming that the total volume of the substance didn't change during the melting-solidification cycles.

3.1.2. Aromatic derivatives thin films and heterostructures preparation by vacuum evaporation

Thin films of PTCDA, Alq3 and TPyP have been prepared by vacuum evaporation and deposition on different substrates (glass/ITO, quartz, Si), which have been cleaned in acetone (glass/ITO, quartz) and with acetone, hydrofluoric acid and distillate water (Si).

Stable, homogeneous organic films, with good adhesion to the substrates have been prepared, by the evaporation of the organic powder contained in the quartz crucible heated by a self-sustaining kanthal winding, in an Alcatel system with turbo molecular pump [40; 46]. During the deposition with a duration between 10-15 min, the temperature was measured by a thermocouple situated at the bottom of the crucible and varied between 220-240 °C for PTCDA [40], 150-160 °C for Alq3 [40] and 175-185 °C for TPyP [46].

3.1.3. Aromatic derivatives thin films an heterostructures preparation by MAPLE

A special type of Pulsed laser Deposition (PLD) technique, Matrix Assisted Pulsed Laser Evaporation (MAPLE), has been used for the deposition of small molecule organic films (PTCDA, ZnPc, Alq3). This technique involves the ablation of a target formed by the frozen solution of the organic compound in a high molecular weight and strong laser wavelength absorbing solvent, like dimethylsulphoxide (DMSO) or chloroform. Deposition was realized with a KrF* laser, Coherent ComplexPro 205 characterised by λ=248 nm, τ_{FWHM}~25 ns, repetition rate=10 Hz [47]. The incident laser energy absorbed by the solvent molecules is converted into thermal energy determining the heating and simultaneous evaporation of the two components. The solvent molecules are pumped away by the vacuum pump that maintains a pressure of 10^{-2}-10^{-1} mbar in the deposition chamber, while the less volatile molecules of the organic compound deposit on the substrate maintained at room temperature. The low value fluence varied between 160 mJ/cm^2 and 430 mJ/cm^2 to avoid the deterioration of the organic molecule and the number of pulses between 10000 and 120000, with effect on the films' thickness (40 nm-150 nm), which has been evaluated by ellipsometry.

3.2. Optical properties of aromatic derivatives thin films

The absorption spectrum of m-DNB, presented in Figure 20, which is similar to a classical semiconductor, could be correlated with the strong interactions between the polar molecules and with the partial superposition of the π-electrons clouds from neighbour molecules generating narrower valence and conduction bands. This spectrum is different from that of benzil presented in Figure 19, which is characterized by a two edges of the fundamental absorption, with a subband light absorption peak situated at 380 nm, attributed to absorption by dicarbonyl groups, strongly interacting in the solid state and producing the split of the energetic level (n, π^*).

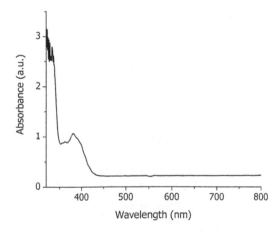

Figure 19. Absorption spectra of benzil film grown between two quartz plates [45].

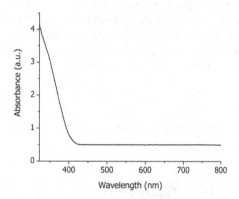

Figure 20. Absorption spectra of m-DNB film grown between two quartz plates [45].

From Figure 21 it can be emphasized that the shape of the fundamental absorption edge is not affected by the presence of impurities. No important changes have been evidenced at the absorption edge characterized by a lower energetic threshold. But the absorption at the edge characterized by the higher energetic threshold is attenuated in benzil doped with m-DNB or sodium compare to pure benzil because of the light scattering process on the nonhomogeneities of the films. This effect is stronger in benzil doped with Na because it is not completely dissolved, segregates and generates microinclusions as a distinct phase.

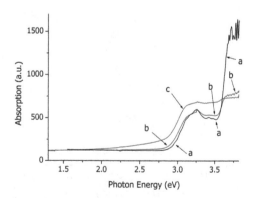

Figure 21. Comparative absorption spectra of: pure benzil (a); benzil doped with m-DNB (3 wt %) (b); benzil doped with Na (6 wt %) (c): grown between two quartz plates [45].

The effect of the impurities on the shape and position of the absorption peak in benzil situated at 3.25 eV and assigned to dicarbonyl group absorption is not important and no other absorption peaks have been evidenced in the longer wavelength range to sustain the trapping of the excitation energy by theses impurities. Therefore we can conclude that the energetic levels of these impurities are not significantly lower than the lowest energetic level which characterizes the crystalline assembly [21].

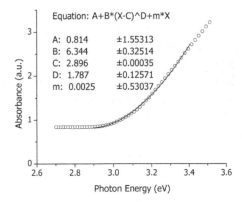

Figure 22. Fitting of the experimental data for pure m-DNB film [45].

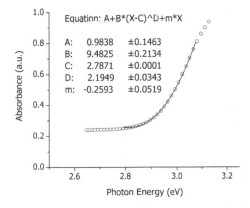

Figure 23. Fitting of the experimental data for pure benzil film in the range 2.6 eV-3.1 eV [45].

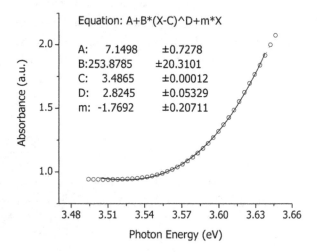

Figure 24. Fitting of the experimental data for pure benzil film in the range 3.49 eV-3.65 eV [45].

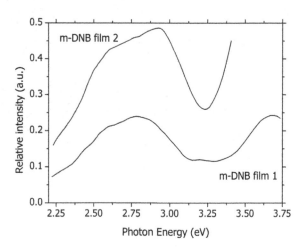

Figure 25. Emission spectra of pure m-DNB films of different thickness for $\lambda_{excitation}$=300 nm [21].

The band gap energy has been evaluated from the experimental data fitting, near the fundamental absorption edge, shown in Figure 22 for m-DNB and, Figure 23 and Figure 24 for benzil, using the formula (4) and we have obtained E_g=2.90 eV for m-DNB and E_{g1}=2.79 eV and E_{g2}=3.54 eV for the two absorption edges in benzil.

Supplementary information about the optical properties of theses aromatic derivatives in solid state have been obtained from the luminescence measurements. At excitation with λ=300 nm, the emission spectra of pure m-DNB present a peak situated at 2.92 eV with a shoulder at 2.63 eV, as it is shown in Figure 25. The apparent red shift of the emission peak situated at 2.92 eV to 2.78 eV in thicker film is due to a self-absorption process of the emitted radiation and not to a recombination on the energetic levels associated with physical defects or impurities.

Luminescence spectra of benzil films presented in Figure 26 show a peak at 2.30 eV attributed also to the radiative decay from the excited triplet state (T_1) to the ground state (S_0), transition possible because of the vibrational interactions as was mentioned for bulk samples in paragraph 2.2.

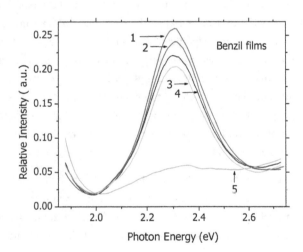

Figure 26. Emission spectra of pure and doped benzil films: (1) benzil doped with Na (1 wt %); (2) benzil doped with m-DNB (3 wt %); (3) benzil doped with Ag (2.4 wt %); (4) benzil pure; (5) benzil doped with Na (6 wt %) [21].

The excited molecules of benzil are more sensitive to a radiationless process and the conversion from the first singlet excited state to the lowest triplet excited state becomes possible by „intersystem crossing", transition between states with different multiplicity, very efficient in systems containing carbonyl groups, like benzil. This radiationless process is followed by a radiative decay through phosphorescence from (T_1) to (S_0). The intersystem crossing appears in benzil as a transition between (n, π^*) states of singlet and triplet levels rather than between (π, π^*) states. Both absorption and emission transitions involve states localized on carbonyl groups, which emit only from planar configuration.

The behaviour of benzil molecule depends on the flexible conformation [48] because the torsion angle around the central bond C-C can change and the geometry of the molecule can change after excitation with light. In the ground state the benzil molecule has a skew configuration and can twist around the carbonyl-carbonyl bound with little interaction between the two benzyl halves of the molecule. This interaction becomes strong in excited molecule, which rearranges in a new configuration with a trans-planar dicarbonyl system characterized by a redistribution of the energy followed by the process of „intersystem crossing". By phosphorescent emission the system passes from a trans-planar configuration to the ground state with also a trans-planar configuration considering the Frank-Condon principle [49]. In the next step the molecule relax from the emissive trans-planar configuration to a skew configuration and the differences in the emission spectra of benzil can be determined by changes in the molecular conformation of the ground state [50]. For most of the dopants we have not remarked any modification in the shape and position of the emission peak that sustains no modification in the molecular conformation of the (n, π^*) state with effect on the angle between the carbonyl groups. The only changes, a slightly blue shift and significant attenuation in intensity, have been remarked in Figure 26 for benzil highly doped with Na (6 wt %). A possible charge transfer from Na atoms to oxygen atoms in carbonyl groups can generate conformational changes, and the shift could be explained by the decrease in the dihedral angle between the two carbonyl groups [21]. The only impurity active in emission is sodium at high concentration because of the conformational changes in the emitting triplet state, while m-DNB, Ag, Na are not active in absorption.

In Figure 27 and Figure 28, we have emphasized the band structure of the absorption spectra in PTCDA and Alq3 with peaks situated at 358 nm, 374 nm, 475 nm, 552 nm and 232 nm, 261 nm, 380 nm respectively. This confirms the presence of median isomer in Alq3 film. The position of the two important peaks situated at 2.25 eV and 2.61 eV remain unchanged in the absorption spectrum of PTCDA deposited on glass covered by ITO while the absorption spectrum of the heterostructure with double organic layer, glass/ITO/PTCDA/Alq3, preserves the pattern of the absorption spectrum of PTCDA between 400 nm and 600 nm, which is very important in the stage of the charge carrier's generation.

In Figure 29 have been evidenced the presence of π-π^* absorption bands characteristic for free-base ethio type porphyrin. These bands are associated with π-π^* transition between bonding and anti-bonding molecular orbitals. Other bands which have been also evidenced are: an intense Soret band (B) with a peak centred at 430 nm and 4 Q bands situated at 520 nm, 555 nm, 590 nm and 645 nm.

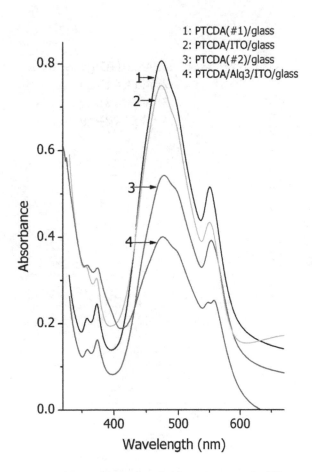

Figure 27. Absorption spectra of PTCDA thin films deposited by vacuum evaporation [40].

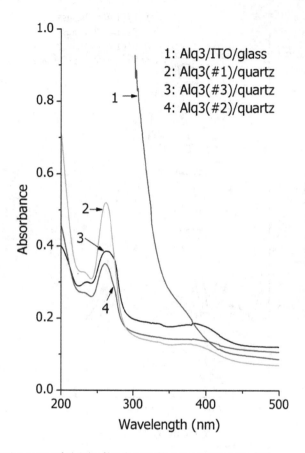

Figure 28. Absorption spectra of Alq3 thin films deposited by vacuum evaporation [40].

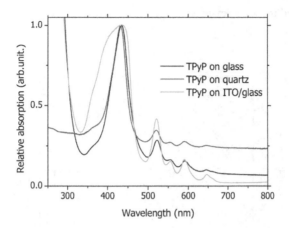

Figure 29. Absorption spectra of TPyP thin films deposited by vacuum evaporation [46].

The shape of the absorption spectra of TPyP thin films deposited on different substrates is preserved at λ> 430 nm. The slight red shift of the Soret band can be determined by the order induced by the interaction between the molecules in solid state influenced by the interaction with the substrate [51]. A possible bonding mechanism can be based on the pyridyl-surface interaction mediating the deformation of the molecule after adsorption on the substrate's surface. Subsequent packing of the molecules can be determined by the non-covalent interactions mediated by the terminal pyridyl groups and these interactions seem to prevail over the site-specific adsorption [52]. During these intermolecular interactions the porphyrin core can be deformed and the symmetry of the TPyP molecule modified because the conformation of this molecule is defined by several degree of freedom (dihedral angle in correlation with the rotation of the pyridyl group about C-C bond, inclination angle of the same bond and distortion angle determined by the steric repulsion between hydrogen atoms of the pyridyl group and the pyrrole moieties [53]).

The spectra in Figure 30 have revealed a wide absorption band situated between 400 nm and 600 nm in PTCDA with a maximum at 480 nm and a shoulder at 550 nm, this shape being determined by the interactions of the π-electrons system of the neighbour planar molecules very closed packed in solid state [21; 54]. The excited states can be the result of the superposition of the intramolecular Frenkel excitons and intermolecular charge transfer excitons existing near the excitation threshold [55]. The UV-VIS spectra of Alq3 confirm that the low temperature isomer (median) correlated with the presence of the weak absorption band situated at 380 nm, dominates in the films deposited at room temperature [56]. For ZnPc we have evidenced two absorption peaks situated at 690 nm, a strong band corresponding to Q band and a weak band situated at 330 nm corresponding to B band [57].

Modifications in the deposition parameters (target concentration, fluence and number of pulses) are reflected in the thickness of the layer and not in the shape of the transmission.

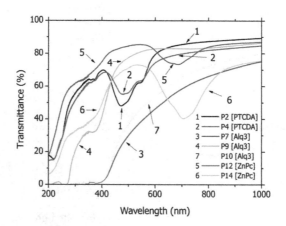

Figure 30. UV-VIS spectra of organic thin films deposited by MAPLE: PTCDA on quartz (1 and 2); PTCDA on ITO (7); Alq3 on quartz (3 and 4); ZnPc on quartz (5 and 6) [54].

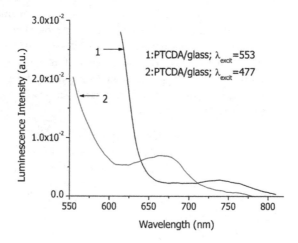

Figure 31. Emission spectra of PTCDA film deposited by vacuum evaporation on glass for two excitation wavelengths [40].

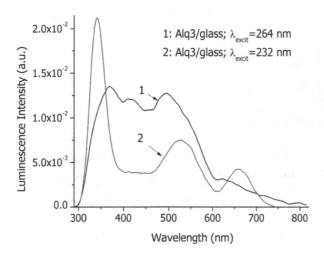

Figure 32. Emission spectra of Alq3 film deposited by vacuum evaporation on glass for two excitation wavelength [40].

The fluorescence emission spectrum of PTCDA film presented in Figure 31 shows a broad structureless band shifted significantly to the red compared to the PTCDA fluorescence spectrum in solution as a consequence of a strong interaction between the organic molecules in the solid state favoured by the close spacing and important overlap of the molecular planes. The emission takes place in PTCDA from the lowest excited singlet state (S_1) by the relaxation of the electron transferred by the light absorption on an antibonding π orbital [40]. The preservation of the emission peak situated at λ=680 nm for two different excitation wavelengths sustains the compositional homogeneity of the film. The luminescence in Alq3, presented in Figure 32, is generated by excitations localized on individual molecule with optical properties independent of molecular environment [58]. The presence of two isomers with different spatial configurations is sustained by the generation of different emission peaks for different excitation wavelengths. The significant Stocks shift in the both spectra of PTCDA (ΔE=0.40 eV) and Alq3 (ΔE=0.9 eV) between the peaks of the lowest level absorption and highest fluorescence emission level, and large Frank-Condon shift (0.40-2.3 eV) measured peak to peak between the absorption and emission spectra, can be correlated with effects determined by the solid state structure and with important conformational differences between the ground state and the excited state [40].

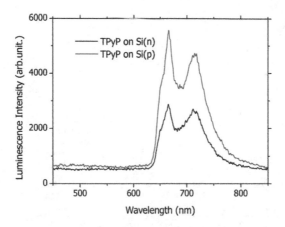

Figure 33. Photoluminescence spectra of TPyP films deposited by vacuum evaporation on different types of Si substrates [46].

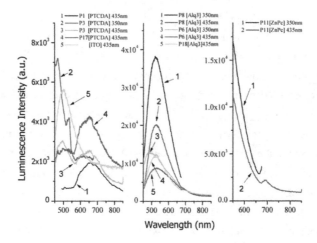

Figure 34. Photoluminescence spectra of organic thin films deposited by MAPLE for two excitation wavelengths: (a) PTCDA on Si (1 and 2), PTCDA on ITO/ZnPc (4); (b) Alq3 on Si (1 and 3), Alq3 on ITO/ZnPc (5); (c) ZnPc on Si (1) [54].

The photoluminescence spectra of TPyP films deposited on Si by vacuum evaporation have revealed in Figure 33 two emission bands situated at 660 nm and 700 nm associated to Q bands and corresponding to free-base ethio type porphyrin [46].

The photoluminescence investigations have evidenced the preservation of the chemical structure of the compounds (PTCDA, Alq3, ZnPc) during the deposition by MAPLE, because we have identified the characteristic emission peaks corresponding to each compound, as can be seen in Figure 34. The emission peak situated at 500 nm, in PTCDA deposited on Si, is associated with monomer-like species and that situated at 650 nm can be associated with two excimeric states [54]. The emission peak in Alq3 film deposited on Si is situated at 520 nm being associated with the excitation of median isomer dominant at the deposition temperature. The emission peak in ZnPc film deposited on Si is situated between 650 nm and 750 nm and can be associated with the deexcitation from the first excited singlet state with an energy of 1.8 eV [54]. In the heterostructures with double organic layer the weak emission of ZnPc is masked by the stronger emission of PTCDA and Alq3.

3.3. Electrical properties of aromatic derivatives thin films

A material could become interesting for optoelectronic application if it is adequate from the point of view of both optical and electrical properties (such as good contact inject and transport the charge carriers). A good injection of the charge carrier was evidenced at the ITO/m-DNB contact compared to ITO/benzil contact as can be shown in Figure 35. This can be explain by the difference in the energetic contact barrier, which is higher between ITO and benzil than ITO and m-DNB, as a consequence of the position of the highest occupied molecular orbital (HOMO) and lowest unoccupied molecular orbital (LUMO) in these aromatic derivative compounds.

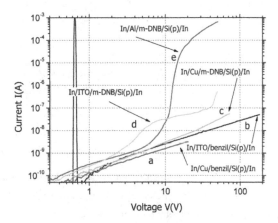

Figure 35. I-V characteristics of the benzil and m-DNB based heterostructures with Si(p) anode and different cathode (Cu, ITO, Al): (a) Cu for benzil ; (b) ITO for benzil; (c) Cu for m-DNB; (d) ITO for m-DNB, (e) Al for m-DNB [61]. For contacting we have used very high purity In.

For the sample based on m-DNB two different regions were identified on the I-V characteristic: the ohmic behaviour region at low voltages and a region with a behaviour associated
to the space charge limited effect at voltages > 5 V. Both in Figure 35 and Figure 36 can be
seen that at voltages > 5 V the effect of the space charge limitation of the current becomes
important in the heterostructure ITO/m-DNB/Si(p). The steep increase in the current at voltage ~ 10 V for Al/m-DNB/Si(p) and ITO/m-DNB/Si(p) can be associated with an avalanche
generation mechanism involving energetic states situated in the band gap, in the interface
region, state generated by the easy diffusion of Al in organic layer favoured by the first ioni-
zation potential of Al (5.98 eV).

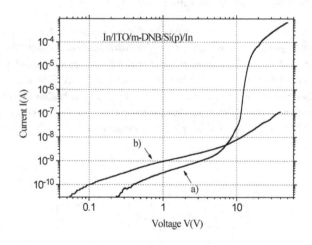

Figure 36. I-V characteristics for ITO/m-DNB/Si(p) heterostructures for different type ITO contacts. For contacting we
have used very high purity In [61].

The different ITO/m-DNB/Si heterostructures have shown significantly different shapes of
the I-V characteristics as a consequence of the crystalline quality of the organic layer, in correlation with the preparation method (rapid thermal directional solidification).

A blocking diode behaviour, both at direct and reverse bias, has been emphasized in Figure
37 at low applied voltages in ITO/TPyP/Si heterostructures, independent of the type of conduction of the Si electrode. For these heterostructures no photoelectric effect has been evidenced. These I-V characteristics are quasi-linear at low voltages and at higher voltages the
limitation of the current determined by the space charge and/or by trap-charge became very
important as can be seen in Figure 37.

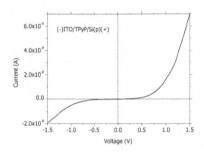
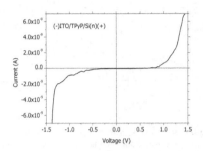

Figure 37. I-V characteristics for ITO/TPyP/Si heterostructures [46].

The heterostructure Au/TPyP/Si, at an illumination through the metallic electrode and direct polarization, shows a rectifier behaviour presented in Figure 38, determined by the energetic barrier at the contact Au/TPyP, which can be lowered applying a voltage > 0.30 eV. The linear behaviour at low applied voltages became a power dependence with n>2 at voltages >0.1 V and corresponds to trap charge limited current. At reverse bias, the same heterostructure shows a blocking behaviour independent of the applied voltage, because the energetic barrier is too high and can't be surpass by the charge carriers [46].

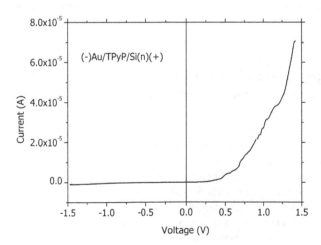

Figure 38. I-V characteristic of the Au/TPyP/Si heterostructure [46].

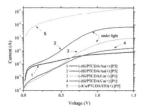 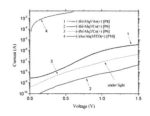 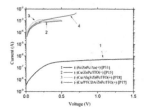

Figure 39. I-V characteristics of Si or ITO/organic layer(s)/Au or Cu heterostructures prepared by MAPLE: based on PTCDA (a) curves 2 and 3 under illumination; based on Alq3 (b) curve 3 under illumination; based on ZnPc (c) all in dark [54].

In Figure 39 a, b, c are presented the I-V characteristics in dark and under illumination, the highest current (~10^{-4}-10^{-3} A) being obtained in dark with the structure prepared with PTCDA, Alq3 or ZnPc on ITO substrate, at low applied voltage of 0.5 V [54]. This current is with three orders higher than the current for the same structure realized on Si. This behaviour is correlated, in the first case, with the height of the energetic barriers at the interfaces that favour the injection of holes from ITO positively biased in organic. The I-V characteristics obtained under continuous illumination at an applied voltage of 1 V, indicate a higher current in the heterostructures realized with PTCDA and, Si and Cu electrodes, explained by the higher energetic barrier for electron injection at the contact Alq3/Cu (ΔE=1.5 eV) compared to PTCDA/Cu (ΔE=0.7 eV). The current is one order higher than the dark current confirming the photo generation process [54]. In the heterostructures with double organic layer and, ITO and Cu electrodes, we have obtained a current of 2x10^{-3} A at 0.5 V, explained by the energetic barrier in ITO/ZnPc/Alq3/Cu heterostructure and by the presence of the interface dipoles reducing the energetic barrier and improving the conduction in ITO/ZnPc/ PTCDA/Cu heterostructure.

4. Organic/organic composite films based on aromatic derivative inclusions for optoelectronic applications

Lately, a special attention has been paid to composite materials based on different organic polymeric matrix and organic inclusions to obtain materials combining the properties of the both components [25; 34; 45; 62; 63]. This field of research has developed from fundamental investigations to the synthesis of new monomers to be introduced in polymeric matrix. The most important advantage of the polymeric matrix is the possibility to deposit thin films using inexpensive methods, such as the deposition from solution by spin-coating. The limiting parameter is the quality of the layer and can be controlled by the control of the experimental conditions. A special attention is focused to the identification and development of π-conjugated systems with functional groups that assure an improvement in the emission properties and charge carrier mobility necessary for optoelectronic applications.

We have emphasised the effect of the polycarbonate of bisphenol A matrix on the properties of the synthesised amidic monomers with –CN and –NO$_2$ substituent groups with the purpose to manipulate the local molecular environment of the monomer for changing the physical properties of the films (transmission, luminescence, electrical transport) in correlation with the quality of the spin-coated layers.

The polycarbonate of bisphenol A, utilized as matrix, is characterised by a large domain of transparency, high transmission in visible, high refraction index, solubility in common solvents. As inclusions, to be embedded in the matrix, we have selected monomers characterised by a maleamic acid structure with different functional groups:

$$\text{HOOC}-\underset{}{\overset{H}{C}}=\underset{}{\overset{H}{C}}-\underset{\underset{O}{\parallel}}{C}-R-\overset{R_2}{\underset{}{\bigcirc}}-R_1$$

where R=-NH, R$_1$=-CN for (MM3); R=-NH, R$_1$=-NO$_2$; R$_2$=-NO$_2$ for (MM5) [64].

After testing the process of layer formation in correlation with the surface energy by contact angle measurements using two different solvents, we have selected dimethylformamide (DMF) for the preparation of the "mother solution" that contain the both components, matrix and inclusion. We have varied the weight ratio between the components 1/3;1/2; 1/1, using the pre-wetting of the surface and different duration and rotation speeds for the spreading stage (t$_1$=3s; 6s; 9s; 12s; v$_1$=0.5 krpm; 0.7 krpm; 0.9 krpm; 1.13 krpm) and homogenisation stages (t$_2$= 10s; 20s; v$_2$=1.6 krpm; 1.9 krpm; 2.2 krpm; 2.7 krpm; 3 krpm), with the purpose to identify the most adequate conditions for the deposition of layers [64].

UV-Vis transmission spectra presented in Figure 40 have evidenced differences in the behaviour of the composite material prepared with (MM3) and (MM5), determined by differences in the chemical structure of these components. The shape of the transmission curve is determined by the substituent to the aromatic nucleus and depends on the lone electron pairs of the oxygen atoms in the carbonyl and nitrous groups involving (n, π^*) state, which are splitted because of the interaction in the solid state between the polycarbonate matrix and (MM5) monomer. No significant difference has been emphasised, in the UV-VIS spectra, between the monomer deposited by vacuum evaporation and the same monomer embedded in a polymeric matrix and deposited by spin-coating. Although (MM3) shows also lone electron pairs the interaction between the cyan groups and the carbonyl groups is not so intense in the solid state to favour the splitting of the (n, π^*) energetic level.

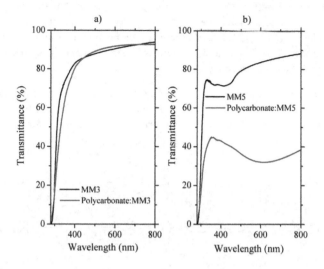

Figure 40. Comparative UV-VIS spectra of MM3 and MM5 monomers deposited by vacuum evaporation and polycarbonate /MM3 and polycarbonate /MM5 deposited by spin coating, on glass substrate [64].

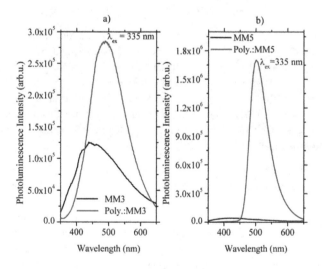

Figure 41. Photoluminescence spectra of monomers (MM3) and (MM5) deposited by vacuum evaporation and polycarbonate/MM3 and polycarbonate/MM5 deposited by spin coating on glass substrates [64].

The emission of the polymer/monomer composite material is determined by the interaction between the chromophoric groups in monomer with the polymeric matrix. This interaction can generate the shift, broadening or strengthening of the emission peak [64].

The Figure 41 shows that the polymeric matrix significantly affects the emission spectra of the monomers characterised by a peak situated at 430 nm in (MM5) and 450 nm in (MM3). These shapes of the spectra can be correlated with the emission properties of the substituted benzene nucleus [65] and with the involvement of the (n, π^*) states lower than the usual singlet excited states. The strongest emission was obtained for monomer (MM5) in polycarbonate of bisphenol A matrix probably due to the strong absorption of the excitation radiation (λ=335 nm) assuring a higher efficiency of the emission process.

In composite material based on (MM5) the emission spectra show a maximum around 510 nm and in composite material based on (MM3) a slightly weaker and broader maximum around 480 nm. Figure 41 has not evidenced a strong broadening effect of the matrix on the emission spectrum of monomers. The emission of polycarbonate:MM5 is not blue shifted and therefore we suppose that the monomer is not highly stressed in the polycarbonate matrix.

In Figure 42 are presented some results on the investigation of the effect of the polymeric matrix on the electrical transport properties of Si/monomer/Si and Si/polycarbonate:monomer/Si heterostructures.

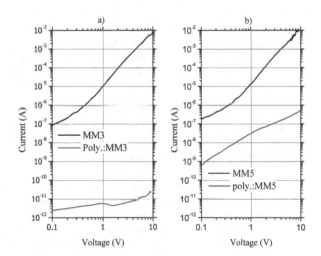

Figure 42. I-V characteristics of Si/MM3/Si and Si/polycarbonate:MM3/Si heterostructures (a) and Si/MM5/Si and Si/polycarbonate/MM5/Si heterostructures (b) [64].

We have analysed the electrical properties of the heterostructure Si/monomer/Si and Si/polycarbonate:monomer/Si at room temperature, testing the reproducibility of the measurements and considering only the typical behaviour. The heterostructures realized with composite materials show a current with 3 orders lower than the heterostructures realized only with monomers. The charge carrier transport is mainly affected by the insulating character of the polymeric matrix. The highest current (3×10^{-8} A) has been obtained in heterostructure Si/polycarbonate:MM5/Si for an applied voltage of 1 V and for voltages between 0.1 V and 1 V the characteristic is weakly superlinear [61]. For the heterostructures realized only with monomers the I-V characteristics are linear at low voltages and become strongly superlinear for voltages >0.2 V.

The films obtained from the polycarbonate containing the monomer with two nitrous substituents (MM5) to the aromatic nucleus have shown good transparency, and photoluminescence in the green region and promising electrical properties at voltages >0.6 V ($I = 10^{-8}$ A) with a close to linear characteristic at voltages between 1 V and 10 V [64].

Also this organic/organic composite material seems to be promising for optoelectronic applications, the spin coated composite layers are characterised by a specific morphology and a high degree of disorder which affect the optical and electrical properties and make difficult their control.

5. Conclusions

In this chapter we summarize some of the most important results of our work in the field of new materials for applications in the field of optoelectronics. Our interest was focused on organic molecules containing electrons, occupying non-localised molecular orbitals in strongly conjugated systems, such as aromatic derivatives compounds (benzil, m-DNB, PTCDA, ZnPc, Alq3, TPyP) for which we have evidenced large transparency domain and good fluorescence emission.

We have realised a comparative investigation on the properties of the same aromatic derivative compound as bulk and thin film material showing both good optical, including luminescent, properties. The interest in studying bulk organic crystals is justified by the perspective to use these materials as a crystalline host matrix both for organic and inorganic guests/inclusions. The organic matrix assures an efficient fluorescence mechanism and from the guest component it is expected an improvement in stability, emission properties of the matrix and electrical mobility. A special attention was paid to the preparation methods both for bulk crystals (emphasising the correlation between the growth interface stability and quality of the organic crystal) and thin films (emphasising the effect of the thin film deposition method -directional solidification, vacuum evaporation, MAPLE- on the properties of the organic film and heterostructures).

Thin films from the above mentioned aromatic derivatives are preferred as organic matrix for host/guest systems because of the major problems associated with bulk crystalline matrices determined by the difficulties of their growing, processing and doping to assure homogeneous distribution of the guest atoms.

Important results have been bought in the field of molecular organic matrix based on aromatic derivatives (benzil, m-DNB)/inorganic (iodine, silver, sodium) or/and organic (m-DNB, naphthalene, bulk) composite systems and in the field of composite films prepared from polymeric matrix and active monomeric inclusions based on π-conjugated systems containing functional groups with special properties, to improve the properties of film forming, emission properties of the matrix and the charge carrier mobility in the matrix, with the purpose to obtain materials for potential optoelectronic applications.

We have emphasised the effect of the polycarbonate of bisphenol A matrix on the properties of the synthesised amidic monomers with –CN and $-NO_2$ substituent groups with the purpose to modify the local molecular environment of the monomer and change the optical and electrical properties of the films.

Author details

Florin Stanculescu[1*] and Anca Stanculescu[2]

*Address all correspondence to: fstanculescu@fpce1.fizica.unibuc.ro

1 University of Bucharest, Bucharest-Magurele,, Romania

2 National Institute of Materials Physics, Bucharest-Magurele,, Romania

References

[1] Helfrich, W., & Schneider, W. G. (1965). Recombination radiation in anthracene crystals. *Phys Rev. Lett.*, 14(7), 229-231.

[2] Tang, C. W., & Van Slyke, S. A. (1987). Organic electroluminescent diodes. Appl. Phys. Lett. , 51(12), 913-915.

[3] Van Slyke, S. A., Chen, C. H, & Tang, C. W. (1996). Organic electroluminescent devices with improved stability. *Appl. Phys. Lett.*, 69(15), 2162-2160.

[4] Kijima, Y., Asai, N., & Tamura, S. I. (1999). A blue organic light emitting diode. Jpn. J. Appl. Phys.Part 1 , 38(9A), 5274-5277.

[5] Hung, L. S., Tang, C. W., & Mason, M. G. (1997). Enhanced electron injection in organic electroluminescence devices using an Al/LiF electrode. Appl. Phys. Lett. , 70(2), 152-154.

[6] Parthasarathy, G., Shen, C., Kahn, A., & Forrest, S. R. (2001). Lithium doping of semiconducting organic charge transport materials. *J. Appl. Phys.*, 89(9), 4986-4992.

[7]   Brütting, W., Riel, H., Beierlein, T., & Riess, W. (2001). Influence of trapped and inter-facial charges in organic multilayer light-emitting devices. J. Appl. Phys. , 89(3), 1704-1712.

[8]   Xue, J, & Forrest, S. R. (2004). Carrier transport in multilayer organic photodetectors: I. Effects of layer structure on dark current and photoresponse. *J. Appl. Phys.*, 95(4), 1859-1868.

[9]   Huang, J., Blochwitz-Nimoth, J., Pfeiffer, J. M., & Leo, K. (2003). Influence of the thickness and doping of the emission layer on the performance of organic light-emitting diodes with PiN structure. J. Appl. Phys. , 93(2), 838-844.

[10] Parthasarathy, G, Burrows, P. E, Khalfin, V, Kozlov, V. G, & Forrest, S. R. (1998). A metal-free cathode for organic semiconductor devices. *Appl. Phys. Lett.*, 72(17), 2138-2140.

[11] Gu, G., Parthasarathy, G., & Forrest, S. R. (1999). A metal-free, full-color stacked organic light-emitting device. Appl. Phys. Lett. , 74(2), 305-307.

[12] Ichikawa, M., Amagai, J., Horiba, Y., & Koyama, T. (2003). Dynamic turn-on behavior of organic light-emitting devices with different work function cathode metals under fast pulse excitation. J. Appl. Phys. , 94(12), 7796-7800.

[13] Guo, T. F., Yang, F. S., Tsai, Z. J., & Feng, G. W. (2006). High-brightness top-emissive polymer light-emitting diodes utilizing organic oxide/Al/Ag composite cathode. Appl. Phys. Lett. 051103-1-051103-3 , 89(5)

[14] Barth, S., Müller, P., Riel, H., Seidler, P. F., Rieß, Vestweber. H., & Bässler, H. (2001). Electron mobility in tris(8-hydroxy-quinoline)aluminium thin films determined via transient electroluminescence from single- and multilayer organic light-emitting diodes. J. Appl. Phys. , 89(7), 3711-3719.

[15] Kalinowski, J, Picciolo, L. C, Murata, H, & Kafafi, Z. H. 2001. Effect of emitter disorder on the recombination zone and the quantum yield of organic electroluminescent diodes. *J. Appl. Phys.*, 89(3), 1866-1874.

[16] Sun, J. X., Zhu, X. L., Meng, Z. G., Yu, X. M., Wong, M., & Kwok, H. S. (2006). An efficient stacked OLED with double-sided light emission SID International Symposium. *Digest of Technical Papers SID Symposium Digest*, 37, 1193-1196.

[17] Shao, Y., & Yang, Y. (2005). Organic solid solutions: formation and applications in organic light-emitting diodes. *Adv. Funct. Mater*, 15(12), 1781-1786.

[18] Burroughes, J. H., Bradley, D. C., Brown, A. R., Marks, R. N., Mackay, K., Friend, R. H., Burn, P. L., & Holmes, A. B. (1990). Light-emitting diodes based on conjugated polymers. *Nature*, 347(6293), 539-541.

[19] Tse, S. C., Tsung, K. K., & So, S. K. (2007). Single-layer organic light-emitting diodes using naphthyl diamine. Appl. Phys. Lett. , 90(21), 213502.

[20] Lu, M. H., & Sturm, J. C. (2002). Optimization of external coupling and light emission in organic light-emitting devices: modeling and experiment. J. Appl. Phys. , 91(2), 595-604.

[21] Stanculescu, A., Antohe, S., Alexandru, H. V., Tugulea, L., Stanculescu, F., & Socol, M. (2004). Effect of dopant on the intrinsic properties of some multifunctional aromatic compounds films for target applications. *Synthetic Metals*, 147(1-3), 215-220.

[22] Stanculescu, A., & Popina, Al. (1996). M -Dinitrobenzene optical nonlinear organic crystals growth for optoelectronics. *SPIE Proc*, 2700.

[23] Stanculescu, A., Stanculescu, F., & Alexandru, H. (1999). Melt growth and characterization of pure and doped meta-dinitrobenzene crystals. *J. Cryst. Growth*, 198/199(1-4), 572-577.

[24] Stanculescu, A. (2007). Investigation of the growth process of organic/inorganic doped aromatic derivatives crystals. J. Optoelectron. Adv. Mater. , 9(5), 1329-1336.

[25] Stanculescu, A., Tugulea, L., Alexandru, H. V., Stanculescu, F., & Socol, M. (2005). Molecular organic crystalline matrix for hybrid organic-inorganic (nano) composite materials. *J. Cryst. Growth*, 275(1-2), e1779-e1786.

[26] Sekerka, R. F. (1968). Morphological Stability. *J. Cryst. Growth*, 3(4), 71-81.

[27] Mullins, W. W., & Sekerka, R. F. (1963). Morphological Stability of a Particle Growing by Diffusion or Heat Flow. J. Appl. Phys. , 34(2), 323-329.

[28] Mullins, W. W., & Sekerka, R. F. (1964). Stability of a Planar Interface During Solidification of a Dilute Binary Alloy. J. Appl. Phys. , 35(2), 444-451.

[29] Flemings, M. C. (1974). Solidification processing. *Mc.Graw-Hill, Inc.*

[30] Kitaigorodski, A. I. (1966). *J. Chem. Phys. Phys. Chem Biol.*, 63, 9-16.

[31] Kitaigorodski, A. I. (1973). Molecular crystals and molecules. *Academic Press, New York.*

[32] Halfpenny, P. J., Ristic, R. I., Shepherd, E. E. A., & Sherwood, J. N. (1993). The crystal growth behaviour of the organic non-linear optical material 2-(α-methylbenzylamino)-5nitropyridine (MBANP). *J. Cryst. Growth*, 128(1-4), 970-975.

[33] Leonard, N. J., & Blout, E. R. (1950). The Ultraviolet Absorption Spectra of Hindered Benzils. J. Am. Chem. Soc. , 72(484), 487.

[34] Stanculescu, A., Mihut, L., Stanculescu, F., & Alexandru, H. (2008). Investigation of emission properties of doped aromatic derivative organic semiconductor crystals. *J. Cryst. Growth*, 310(7-9), 1765-1771.

[35] Windsor, M. W. (1965). Luminescence and energy transfer in: D. Fox, M. M. Labes, A. Weissberger (eds.). *Physics and chemistry of the organic solid state, vol. II, Interscience, New York*, 343.

[36] Rothe, C., Guentner, R., Scherf, U., & Monkman, A. P. (2001). Trap influenced properties of the delayed luminescence in thin solid films of the conjugated polymer poly(9,9- di(ethylhexyl)fluorene). J. Chem. Phys. , 115(20), 9557-9562.

[37] Bardeen, J., & Brattain, W. H. (1948). The transistor, a semi-conductor triode. Phys. Rev. , 74(2), 230-231.

[38] Schockley, W. (1949). The theory of p-n junctions in semiconductors and p-n junction transistors. *Bell System Technical Journal*, 28(3), 435-489.

[39] Gaffo, L., Cordeiro, M. R., Freitas, A. R., Moreira, W. C., Gitotto, E. M., & Zucolotto, V. (2010). The effects of temperature on the molecular orientation of zinc phthalocyanine films. J. Mater. Sci. , 45(5), 1366-1370.

[40] Stanculescu, A., Stanculescu, F., Tugulea, L., & Socol, M. (2006). Optical properties of 3,4,9,10-perylenetetracarboxylic dianhydride and 8 -hyfroxyquinoline aluminium salt films prepared by vacuum deposition. *Mat. Sci. Forum*, 514-516, Part 2, 956-960.

[41] Wu, C. I., Hirose, Y., Sirringhaus, H., & Kahn, A. (1997). Electron-hole interaction energy in the organic molecular semiconductor PTCDA. Chem. Phys. Lett. , 272(1-2), 43-47.

[42] Zugang, L., & Nazarè, H. (2000). White organic light-emitting diodes from both hole and electron transport layers. *Synthetic Metals*, 111-112, 47-51.

[43] Curioni, A., & Andreoni, W. (2001). Computer simulations for organic light-emitting diodes. *IBM J of Research and Development*, 45(1), 101-114.

[44] Auwärter, W., Weber-Bargioni, A., Riemann, A., Schiffrin, A., Gröning, O., fasel, R., & Barth, J. V. (2006). Self-assembly and conformation of Tetra-pyridil-porphyrin molecules on Ag(111). *J. Chem. Phys.*, 124(19), 194708-1-194708-6.

[45] Stanculescu, F., Stanculescu, A., & Socol, M. (2006). Light absorption in meta-dinitrobenzene and benzyl crystalline films. J. Optoelectron. Adv. Mater. , 8(3), 1053-1056.

[46] Socol, M, Rasoga, O, Stanculescu, F, Girtan, M, & Stanculescu, A. (2010). Effect of the,orphology on the optical and electrical properties of TPyP thin films deposited by vacuum evaporation. *Optoelectronics and Advanced Materials-Rapid Communications*, 4(12), 2032-2038.

[47] Socol, G., Mihailescu, I. N., Albu, A. M., Antohe, S., Stanculescu, F., Stanculescu, A., Mihut, L., Preda, N., Socol, M., & Rasoga, O. (2009). MAPLE prepared polymerc thin films for non-linear optic applications. Appl. Surf. Sci. , 255(10), 5611-5614.

[48] Evans, T. R., & Leermakers, P. A. (1967). Emission spectra and excited-state geometry of α-diketones. J. Am. Chem. Soc. , 89(17), 4380-4382.

[49] Kohler, B. E., & Loda, R. T. (1981). Spectroscopy of the benzil-bibenzyl mixed crystal system: High resolution optical studies. J. Chem. Phys. , 74(1), 18-24.

[50] Coppens, P., Ma, B. Q., Gerlits, O., Zhang, Y., & Kulshrestha, P. (2002). Crystal engineering, solid state spectroscopy and time-resolved diffraction. Cryst. Eng. Commun., 4(54), 302-309.

[51] George, H., & Guo, Q. (2008). Self-assembled two-dimensional supramolecular architectures of zinc porphyrin molecules on mica. J. Phys.: Conference series, 100(5), 052077.

[52] Auwärter, W., Klappenberger, F., Weber-Bargioni, A., Schiffrin, A., Strunskus, T., Wöll, C., Pennec, Y., Riemann, A., & Barth, J. V. (2007). Conformational adaptation and selective adatom capturing of tetrapyridyl-porphyrin molecules on a copper (111) surface. J. Am. Chem. Soc., 129(36), 11279-11285.

[53] Klappenberger, F., Weber-Bargioni, A., Auwärter, W., Marschall, M., Schiffrin, A., & Barth, J. V. (2008). Temperature dependence of conformation, chemical state and metal-direct assembly of tetrapyridyl-porphyrin on Cu (111). J. Chem. Phys., 129(21), 214702-214702.

[54] Stanculescu, A., Socol, M., Socol, G., Mihailescu, I. N., Girtan, M., & Stanculescu, F. (2011). Maple prepared organic heterostructures for photovoltaic applications. Apll. Phys. A, 104(3), 921-928.

[55] Sadrai, M., Hadel, L., Sauers, R. R., Husain, S., Krogh-Jespersen, K., Westbrook, J. D., & Bird, G. R. (1992). Lasing action in a family of perylene derivatives: singlet absorption and emission spectra, triplet absorption and oxygen quenching constants and molecular mechanics and semiempirical molecular orbital calculations. J. Phys. Chem., 96(20), 7988-7996.

[56] Jian, Z. A., Luo, Y. Z., Chung, J. M., Tang, S. J., Kuo, M. C., Shen, J. L., Chiu, K. C., Yang, C. S., Chou, W. C., Dai, C. F., & Yeh, J. M. (2007). Effect of isomeric transformation on characteristics of Alq3 amorphous layers prepared by vacuum deposition at various substrate temperature. J. Appl. Phys. 123708-1-123707-6, 101(12)

[57] Senthilarasu, S., Sathyamoorthy, R., Latitha, S., Subbarayan, A., & Natarajan, K. (2004). Thermally evaporated ZnPc thin films-band gap dependence on thickness. Sol. Energy Mater. Sol. Cells, 82(1-2), 179-186.

[58] Burrows, P. E., Shen, Z., Mccarty, D. M., Forrest, S. R., Cronin, J. A., & Thompson, M. E. (1996). Relationship between electroluminescence and current transport in organic heterojunction light-emitting devices. J. Appl. Phys., 79(10), 7991-8006.

[59] Ferguson, A. J., & Jones, T. S. (2006). Photophysics of PTCDA and me-PTCDI thin films: effects of growth temperature. J. Phys. Chem.B, 110(13), 6891-6898.

[60] Haas, M., Shi-Xia, L., Kahnt, A., Leiggener, C., Guldi, D. M., Hauser, A., & Decurtins, S. (2007). Photoinduced energy transfer processes within dyads of metallophthalocyanines compactly fused to a Ruthenium (II) polypyridine chromophore. J. Org. Chem., 72(20), 7533-7543.

[61] Stanculescu, A., & Stanculescu, F. (2007). Investigation of the properties of indium tin oxide-organic contacts for optoelectronic applications. *Thin Solid Films*, 515(24), 8733-8737.

[62] Lee, K. J, Oh, J. H, Kim, Y, & Jang, J. (2006). Fabrication of photoluminescent-dye embedded poly(methyl methacrylate) nanofibers and their fluorescence resonance energy transfer properties. *Adv. Mater*, 18(17), 2216-2219.

[63] Koratkar, N. A., Suhr, J., Johsi, A., Kane, R., Schadler, L., Ajayan, P. M., & Bertolucci, S. (2005). Characterizing energy dissipation in single-walled carbon nanotube polycarbonate composite. *Appl. Phys. Lett*, 87(6), 06312-1-06312-3.

[64] Stanculescu, F., Stanculescu, A., Girtan, M., Socol, M., & Rasoga, O. (2012). Effect of the morphology on the optical and electrical properties of polycarbonate film doped with aniline derivatives monomers. *Synthetic Metals*, 161(23-24), 2589-2597.

[65] Dulcic, A., & Sauteret, C. (1978). The Regularities Observed in the Second Order Hyperpolarizabilities of Variously Disubstituted Benzenes. J. Chem. Phys. , 69(8), 3453-3457.

Chapter 2

Design and Modeling of Optoelectronic Photocurrent Reconfigurable (OPR) Multifunctional Logic Devices (MFLD) as the Universal Circuitry Basis for Advanced Parallel High-Performance Processing

Vladimir G. Krasilenko, Aleksandr I. Nikolskyy and Alexander A. Lazarev

Additional information is available at the end of the chapter

1. Introduction

One of the problems in high speed computing is the limited capabilities of communication links in digital high performance electronic systems. Too slow and too few interconnects between VLSI circuits cause a bottleneck in the communication between processor and memory or, especially in multiprocessor systems, among the processors. Moreover, the problem is getting worse since the increasing integration density of devices like transistors leads to a higher requirement in the number of necessary channels for the off-chip communication. Hence, we are currently in a situation, which is characterized by too few off-chip links and too slow long on-chip lines, what is described as the interconnect crisis in VLSI technology [1]. More than ten years the use of optical interconnects is discussed as an alternative to solve the mentioned problems on interconnect in VLSI technology [2]. A lot of prototypes and demonstrator systems were built to prove the use of optics or optoelectronics for off-chip and on-chip interconnects [3]. The possibilities of current VLSI technology would allow integrating a massively-parallel array processor consisting of a few hundred thousand simple processor elements (PEs) on a chip. Unfortunately it would be a huge problem to arrange several of such PE arrays one after the other in order to realize a highly–parallel superscalar and super-pipelined architecture as well as an efficient coupling to a memory chip. The reason for these difficulties is the not sufficient number of external interconnects to move high data volumes from and to the circuits. In optoelectronic VLSI one tries to solve limitation problem by realizing external interconnects not at the edge of a chip but with arrays of optical detectors and light emitters which send and receive data directly out from the chip area. Honeywell has developed such devices with VCSEL diodes (vertical surface emitting laser diodes) and metal – semiconductor – metal photo-detectors in research project [4].

This allows the realization of stacked 3-D chip architecture in principle. The main problems are not the manufacturing and operating of single devices but the combination of different passive optical elements with active optoelectronic and electronic circuits in one system. This requires sophisticated mounting and alignment techniques which allow low mechanical tolerances and the handling of thermal problems. At present the situation for smart detector circuits is much easier. They can be regarded as a subset of OE-VLSI circuits because they consist only of arrays of photo-detectors with corresponding evaluation circuit for analogue to digital converting. Optical detectors based on PN or PIN photodiodes can be monolithically integrated with digital electronics in silicon what simplifies the design enormously compared with OE-VLSI circuits that in addition contain sender devices realized in GaAs technologies. Furthermore smart detector circuits can be manufactured in nearly every semiconductor fabric. Smart detectors or smart optical sensors show a great application field and market potential. Therefore our approach favors a smart pixel like architecture combining parallel signal detection with parallel signal processing in one circuit. Each pixel has its own PE what guarantees the fastest processing.

The strategic direction of solution of various scientific problems, including the problem of creation of artificial intelligence (AI) systems, human brain simulators, robotics systems, monitoring and control systems, decision-making systems, as well as systems based on artificial neural networks, etc., becomes fast-acting and parallel processing of large2-D arrays of data (up to 1024x1024 and higher) using non-conventional computational systems, corresponding matrix logics (multi-valued, signed-digit, fuzzy logics, continuous, neural-fuzzy and others) and corresponding mathematical apparatus [5-11]. For numerous perspective realizations of optical learning neural networks (NN) with two dimensional structure [5], of recurrent optical NN [6], of the continuous logic equivalency models (CLEM) NN [7-10], the elements of matrix logic are required, and not only of two-valued property, threshold, hybrid but also continuous, neural-fuzzy logics and adequate structure of vector-matrix computational procedures with basic operations of above-mentioned logics. Optic and optoelectronic technologies, methods and principles as well as corresponding element base provide attractive alternative for 2D data processing. These technologies and methods successfully decide problems of parallelism, input-output and interconnections. Advanced non-traditional parallel computing structures and systems, including neural networks, require both parallel processing and parallel information input/output. At the same time there are many new approaches that are based on new logics (neural-fuzzy, multi-valued, continuous etc.). The using of the standard sequential algorithms based on a few operations makes the approaches long-running. But only a few of them [12] can be used for processing of 2D data and perform wide range of needed arithmetic and logic operations). Generalization of scalar two-valued logic on matrix case has led to intensive development of binary images algebra (BIA) [13] and 2D Boolean elements for optic and optoelectronic processors [12-17].

Taking into consideration the above-described approach, consisting in universality, let us recollect some known facts regarding the number of functions. The number of Boolean functions of n variables in algebra of two-valued logic (TVL), which is also Boolean algebra, equals 2^{2^n}. In this TVL there are $N_2 = 2^n$ atoms, which are minterms. Functions of n variables k–valued logic ($k>2$) are reflections $A^n \rightarrow A$, where $A=\{0, 1,... k\text{-}1\}$, and the number of functions equals $N_k = k^{k^n}$. Algebra, formed by set $^\wedge C_u = [0, 1]$ or $^\wedge C_b = [-1, 1]$ is called continuous logic (CL) algebra, and the number of CL functions, as reflections $C_u^n \rightarrow C_u$ depending on the CL algebra can be infinite or finite (the set of reflections is always infinite). CL functions are called only those functions of the set N_N, which are realized by formulas. The number $N_\wedge$ of CL functions in the most developed CL algebra – quasi-Boolean Cleenee algebra ($\Delta = (C_u, \wedge, \vee, -)$), in which any function on any set of arguments takes the value of one of the arguments or its negation, is finite. In this case the number $N_\wedge(n)$ of functions of n arguments increases with increase of n very rapidly [4]: $N_\wedge(0)=2$; $N_\wedge(1)=6$; $N_\wedge(2)=84$; $N_\wedge(3)=43918$.

We would like to draw the attention to the fact, that both natural neurons and their more complex physical and mathematical models suggest discrete-analog and purely analog means for information processing with different level of accuracy, with the possibility of re-arrangement of chosen coding system. This, in its turn, requires corresponding image neuron circuit engineering with programmable logic operations, with transition from analog to discrete processing, to storing etc.

Thus, the search of means aimed at construction of elements, especially universal (at least quasi-universal or multifunctional) with programmable tuning, able to perform not only operations of two-valued logic, but other matrix (multi-valued, continuous, neural-fuzzy, etc.) logic operations is very actual problem [15]. One of promising directions of research in this sphere is the application of time-pulse-coded architectures (TPCA) that were considered in works [18-20]. These architectures were generalized in [11], taking into account basic possible approaches as well as system and mathematical requirements. The time-pulse represen-tation of matrix continuous-logic variables by two-level optic signals not only permits to increase functional possibilities (up to universality), stability to noise, stability and decrease requirements regarding alignment and optical system, but also simplify control circuits and adjustment circuits to required function, operation, and keep untouched the whole meth-odological basis of such universal elements construction, irrespective of valuedness of a log-ic and type of a logic.

But there is another approach based on the use of universal logic elements with the structure of multiple-input multiple-output (MIMO) and time-pulse coding. We call such elements - the elements of picture type (PT). At increase of number of input operands and valuedness of logic (up to continuous) the number of executable functions also increases by the expo-nential law. This property allows simplifying operation algorithms of such universal optoe-lectronic logical elements and hence to raise information processing speed. Most general conceptual approaches to construction of universal picture neural elements and their mathe-matical rationales were presented in paper [11]. But those were only system and structural solutions that is why they require further development and perfection. Mathematical and other theoretical fundamentals of design of matrix multi-functional logical devices with fast acting programmable tuning were considered in paper[19], where expediency of functional basis unification, that is promising for optoelectronic parallel-pipeline systems (OEPS) with command-flow 2D-page (picture) organization [20], necessity in arrays of optic or optoelec-tronic triggers (memory elements) of picture type for storage of information and controlling adjusting operands as well as perspective principles of presentation and coding of multi-val-ued matrix data (spatial, time-pulse and spectral) were shown. Besides, the analysis of vari-ous algebra logics [11, 19, 21-24] for functional systems of switching functions, in spite of their diversity allows us to suggest a very useful idea, in our opinion, that lies in following.

It is possible to create more sophisticated problem-oriented processors, in which the specific time-pulse operands encoding and only elements of two-valued logic are used, which will realize functions of different logics, continuous etc. Taking into account the universality, parallel information processing of the universal elements and the use of only two-valued logic elements for implementation of all other operations the approach is a very promising.

That is why the aim of the given work is to consider the results of design and investigation of optoelectronic smart time-pulse coded photocurrent reconfigurable MFLD as basic com-ponents for 2D-array logic devices for advanced neural networks and optical computers.

2. Design and simulation of two variants of the OPR MFLD base cell

2.1. Picture continuous logic elements (PCLE)

Figure 1 shows the structural diagram of picture neural element (PNE) for computation of all basic matrix-continuous-logic (MCL) operations in matrix quasiBoolean algebra $C=((A,B),^\wedge, \check{},-)$ [11] for which in any set of MCL arguments matrix continuous logic function (MCLF) F takes the value of a subregion of one of the arguments or its supplement. The PE of matrix two-valued logic (MTVL), performing MTVL operations over matrix temporal functions $O^i_{\ t}(t)$ (in point of fact two-valued 2D-operands) realize MCLF over continuous logic variables (CLV) $O^i_{\ t}$. The time-pulse coding of a grayscale picture is shown in Figure 1. As it is seen in Figure 2 at each point of picture output of PNE, MCL can be performed over continuous logic variables (CLV) $O^1_{\ ijT},...O^{n=2}_{\ ijT}$, presented by $t^1_{\ ij}...t^n_{\ ij}$ durations of time pulse signals, during each interval T one of the following operations of CL: min(a,b), max(a,b), mod(a-b), mod($a-b$), complementary$\bar{a}=1-a$, equivalence, etc [10, 11, 23]. The du-ration of MTVL formed at the output and as a result of PNE, signal $f^{NE}_{ij}(t)=f^{NE}_{ij}\left(O^1_{ij}(t),\ O^2_{ij}(t)\right)$, is CL function of input binary temporal variables durations. Thus, as it is seen from Figure 3, almost all basic operations of continuous logic, neural-fuz-zy logic, that are shown in work [21], can be realized with the help of the time-pulse coding of variables $X_1,...,X_n$ and universal (or multifunctional)picture element (UPE) of two-valued logic (TVL). But for that pulse width modulator (PWM) of PT is needed. It is not needed to form contrast-conversion (complementary operand) image for analog picture optic inputs if PWMs PT have complementary outputs.

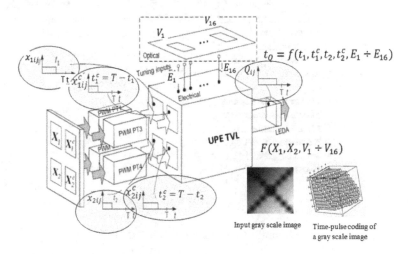

Figure 1. The PNE of matrix-continuous-logic (MCL) with programmable tuning

Thus, becomes obvious that for time – pulse coding realization of PNE of matrix-continuous -logic (MCL) with programmable tuning is necessary UPE of TVL or picture MFLD, by means of which continuously – logic operations over time – pulse signals can be realized. In Figure 1 selection of picture logic functions is carried out by electric adjusting signals and all array cells will realize the same function at the same time. For many appendices it is expedient to choose a logic function at each point of the matrix processor, and therefore there is a desire to make management and tuning also in the form of optical matrix operands. It essentially expands functionality of such processors and MFLD on which basis they are realized.

In work [25] MFLD of two-valued logic (TVL) on current mirrors, photodiodes and LEDs with schemes of their drivers are described and simulated. They are relatively difficult as contain four current mirrors (CM), four schemes XOR, four elements AND and one logic element OR. In the same work different optoelectronic circuitry were offered on base of 2-4 CM and one photo diode, realizing the Boolean operations AND, NOT, OR, NOR, et al with potential and current outputs. They are based on threshold elements, comparators of currents (photocurrents) on current mirrors and circuitry of limited subtraction (CLS). Such base elements also were used for realization of other elements of continuous logic, including operations equivalence (nonequivalence) and etc. [21, 26, 27]. Therefore developing further this approach we use for design of the OPR MFLD.

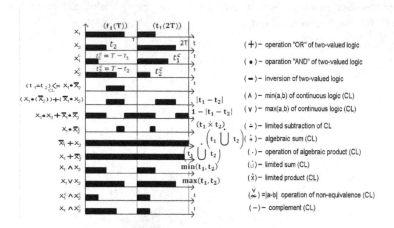

Figure 2. Time diagrams of CL operation fulfillment by means of time-pulse CL variables

2.2. Designing of the base cell for the first version of OPR MFLD-1

The function circuit of the OPR MFLD-1 (the first version) is shown in figure. 3, and the circuit diagram of the OPR MFLD-1 on 1.5μm CMOS transistors is shown in figure. 4. It contains 4 optical inputs (the aperture of photodiodes PD) four cells (PD-CM)$_1$ ÷ (PD-CM)$_4$ executing a role of threshold elements (a threshold -i_0) and realizing operation of the limited subtraction(LS):$i_{CM_i} = i_{pd} - i_0$; current mirror CM5 (or instead of its CM'5 with the optical adjusted threshold) for the reproduction of thresholds$i_0 = 2I_0$; current mirrors CM6-M and CM7-M (M denotes the multiplication currents) for formation together with drivers signals (currents) for four LEDs (2 direct outputs) and LED ' (2 additional inverse outputs).

The cell for the first version of OPR MFLD-1 has a different sub-options, which correspond to different patterns of formation of the thresholdsi_0, namely: 1) sub-option with the formation of all four thresholds using individual current sources, 2) sub-option - with the help of a current mirror - multiplier CM5 and a single current source$2I_0$, 3) sub-option - using the current mirror-multiplier CM'5 with a photodiode for input of the threshold current$i_0 = 2I_0$.

In Figures 5a, 5b it is shown constructive (a matrix fragment – one OPR MFLD-1) the scheme of base nodes and the most simple optical imaging system for connections. The scheme contains 4 photo diodes, 5+8+5=18 transistors (without transistors of drivers) and the scheme is enough simple. By changing optical (or electrical) signals of tuning vector y1÷y4 at input 4 photodiodes signals from light emitter diodes LED and $\overline{LED}$ of the OPR MFLD-1 scheme are moved.

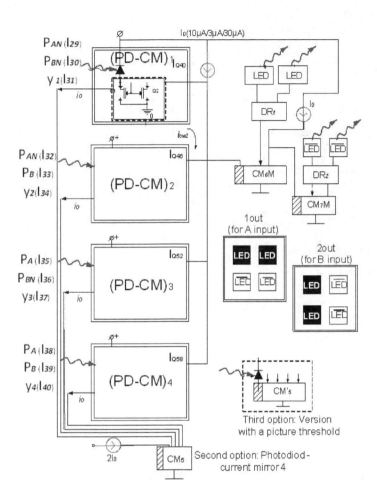

Figure 3. The function circuit of the OPR MFLD-1

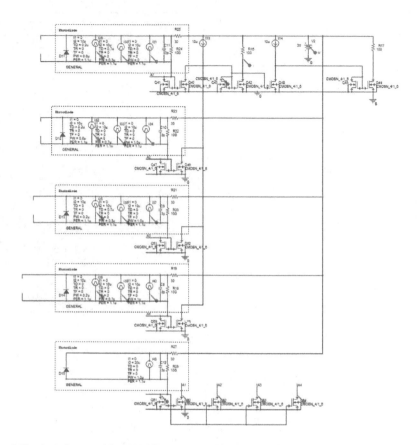

Figure 4. The circuit diagram of the OPR MFLD-1 on 1.5μm CMOS transistors for modeling with OrCAD 16.3 PSpice

Signals from the first input A and from the second input B (a variant of output II) together with tuning vector y1÷y4 will be transformed to a total photocurrent. Base elements of limited subtraction (LS) based on (PD-CM)$_i$ separate out corresponding logic minterms by subtraction of threshold currents i_0 from currents of PDs. We researched various updating of such base circuits. For the task of thresholds it is possible to use the various optical and electric approaches, besides operating generators of currents and various schemes of drivers are possible. A basic accent we nevertheless do on input part of conversion and processing, because forming of matrix of emitters is simpler task, if not to take into account the technological aspects of their integration on a chip.

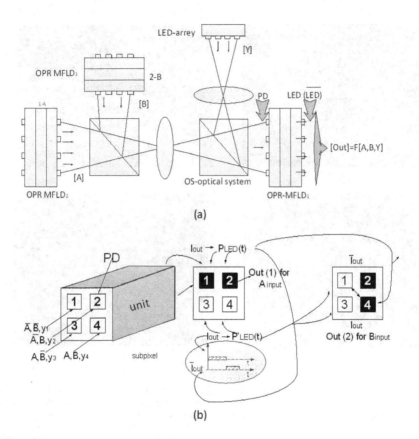

Figure 5. a) Optical plane-to-plane imaging system (b) The constructive scheme of a base cell (fragment) for the OPR MFLD-1

2.3. Simulation of the base cell for the first version of OPR MFLD-1

Results of modeling by means of package OrCAD 16.3 of the offered OPR MFLD-1 are shown in Figure 6 for different tuning signals y1y4 which set necessary functions, for different supply voltage and different amplitudes of currents I_0 (3µA, 10µA and 30µA) accordingly. In Figure 6a, the first diagram above shows the pulses of currents I30, I29, which correspond to the inputs of BN, AN, and current ID (Q40) at the output node (PD-CM)$_1$. On the same Figure, the second, third and fourth diagrams show, respectively, the input and output pulse currents of nodes (PD-CM)$_2$÷ (PD-CM)$_4$. Current pulse I35 duration (see the third diagram in Figure 6a), which is at the input A, equals 2µs. Current pulse I39 duration (see the fourth diagram in Figure 6a) equals 7µs at the input B. The output pulse current ID (Q44) of the circuit is shown in the bottom diagram in Figure 6a and its duration equals 8µS (8=10-2). This confirms the correctness work of the circuit.

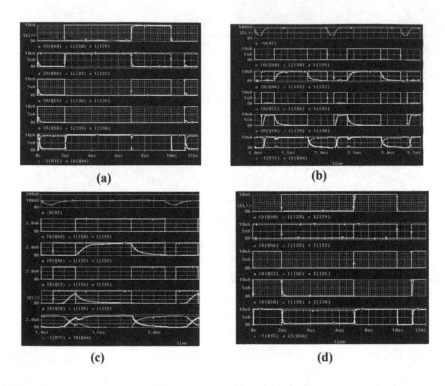

(a)

(b)

(c)

(d)

Figure 6. a) Simulation results of cell of OPR MFLD-1 for functions of NAND two-valued logic(TVL) and NMIN continuous logic (CL) at a supply voltage5 V; (b) Simulation results of cell of OPR MFLD-1 for functions of OR TVL and MAX CL logic (3 V, t(A)=t(I35)=200nS, t(B)=t(I39)=700nS, t=100nS, T=1,1μs); (c) Simulation results of cell of OPR MFLD-1 for functions XOR TVL and NEQ CL (3V, 0.1mW, 3μA, t(A)=t(I35)=200nS, t(B)=t(I39)=700nS,T=1.1μs); (d) Simulation results of cell of OPR MFLD-1 for functions NXOR TVL and EQ CL (3V, 10μA, photo-configurable, t(A)=t(I35)=2μS, t(B)=t(I39)=7μS, t_{out} = t(ID(Q44))=2+3=5 μS)

The diagrams in Figures 6b, 6c, 6d, similar to Figure 6a shows the corresponding input and output currents of the circuit. The difference lies in the different modes for different input pulse durations and the presence of additional power consumption graphics. In Figure 7a dependence of power consumption of OPR MFLD-1 from $I_0$and supply voltage is shown, and in Figure 7b dependence of t_{preset} and $t_{fronts}$from $I_0 = I_{max}$ is shown. From them it is visible, that the power consumption of OPR MFLD-1 P_{drain} (without drivers and output part) is about 0.1-2.5mW. If to take into account that the currents of LEDs must (taking into account the coefficient of transformation and sensitiveness of photo-detectors PD) to be at least in 5÷10 times more, the P_{drain} will increase in 2÷4 times. But, for example, atI_0=10μA, the power consumption will be $P_{drain} \leq 4\div5$mW. At currents 1÷3μA it decreases to 1mW. Delay time is no more than 50÷100 ns, and the period T of time pulse processed signals go into in a microsecond range 1÷16 μs. If to use not 1.5μm technologies CMOS transistors, but more advanced, that is possibility to receive processing time T at level 1÷10ns, i.e. to raise productivity of one channel OPR MFLD-1 to 10^8-10^9 CL-logic operations/sec.

We tested experimentally the circuit for all functions that it can implement. The experiments confirm the implementation of all theoretically possible functions in a wide range of voltages, currents and operating periods of treatment. But given the size limitations of article, here we do not present all results and charts.

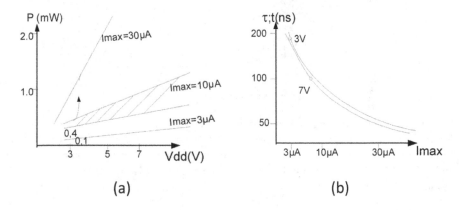

Figure 7. a) Dependence of the power consumption from supply voltage and input current range; (b) dependence of time delay and fronts from supply voltage and input current values

If cells of the MFLD-1 with P_{drain} = 1÷5mW are integrated into array of 32x32 elements or more, the general productivity of such array OPR MFLD-1 will reach 10^{12} CL-logic operations/sec. A modified variant of OPR MFLD-1 in which signals y1,y4 are realized on current generators with possibility of their programming is also offered. Besides, if the array of cells MFLD-1 realizes the same function it is possible to choose signals with sample corresponding nodes (PD-CM)$_i$. The problem of simplification of the optical system is decided in this case. Because it is necessary to give signals not from three optical apertures, but only from two apertures on the OPR MFLD-1 chip.

2.4. Modeling of array of the OPR MFLD-1 with MathCAD

Modeling results of the OPR MFLD-1 with MathCAD which confirm normal functioning of OPR MFLD-1 for all 16 possible functions of binary logic and corresponding functions of continuous logic are shown in figure 8-11. Two inputs 2D operands XA and XB (Figure 8) with dimensional of 32x32 pixels are transformed to XAR and XBR by multiplication of one pixel to 2x2 pixels. Matrixes XAR, XBR have dimensional of 64x64 pixels.

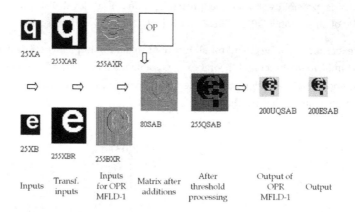

Figure 8. Simulation results of forming and processing processes using OPR MFLD-1

Four matrixes M1÷M4 are formed with formulas shown in Figure 9. These matrixes are used for selection of one subpixel of four pixels of XAR and XBR. Matrixes AXR and BXR are formed after XAR and XBR by elementwise non-equivalence (⊕) operation on matrixes MA and MB. Tuning 2D operand OP is formed by matrixes M1÷M4 and scalar tuning signals oy1÷oy4 or by signals y1÷y4.

$$M1_{i,j} := \mod[(i + 1),2] \cdot \mod[(j + 1),2]$$

$$M2_{i,j} := \mod[(i + 1),2] \cdot \mod[(j + 0),2]$$

$$M3_{i,j} := \mod[(i + 0),2] \cdot \mod[(j + 1),2]$$

$$M4_{i,j} := \mod[(i + 0),2] \cdot \mod[(j + 0),2]$$

$$MA := M1 + M2 \quad OP := \sum_{l=1}^{4} oy_l \cdot M_i ,$$

$$MB := M1 + M3 \text{ where } M_i \in M1 \div M4$$

$$AXR := \overrightarrow{(XAR \oplus MA)} \quad BXR := \overrightarrow{(XBR \oplus MB)}$$

$$SAB := AXR + BXR + OP$$

Figure 9. Transformations formulas for matrixes, tuning operand OP formation and additions

Matrix SAB is formed as sum of AXR, BXR and OP. Threshold processing is done over elements of SAB matrix and matrix QSAB is formed:

$$QSAB_{i,j} = \Phi\left[1 - \Phi\left(3 - SAB_{i,j}\right)\left(3 - SAB_{i,j}\right)\right]\left[1 - \Phi\left(3 - SAB_{i,j}\right)\left(3 - SAB_{i,j}\right)\right] \tag{1}$$

The threshold value tr =3. Four subpixels are united to one pixel with formula

$$UQSAB_{k,l} = QSAB_{2k,2l} + QSAB_{2k,2l+1} + QSAB_{2k+1,2l} + QSAB_{2k+1,2l+1} \tag{2}$$

and output matrix UQSAB dimension is 32x32. Another final threshold processing (t_0=1) is done with formula

$$\text{ESAB}_{k,l} = \Phi\left[1 - \Phi\left(1 - \text{UQSAB}_{k,l}\right)\left(1 - \text{UQSAB}_{k,l}\right)\right]\left[1 - \Phi\left(\text{UQSAB}_{k,l}\right)\left(1 - \text{UQSAB}_{k,l}\right)\right] \qquad (3)$$

and output matrix ESAB is formed.

For more detailed consideration fragments AP, BP, OPP, OSP, OQP, QSP with dimensional of 2x2 subpixels or 4x4 pixels from matrixes AXR, BXR, OP, SAB, UQSAB, ESAB are shown in Figure 10. The fragments are shown as matrixes and images. For conventional presentation of the images in MathCAD the matrixes are multiplied by 80. Output of equivalence operation is QSP with dimensional of 2x2, but for OPR MFLD correct operation matrixes QSAP and QABP with dimensional of 4x4 are used.

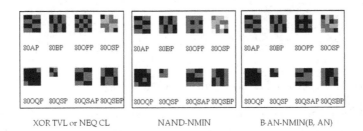

Figure 10. Simulation results of four base cells (2x2 subpixel) of matrix OPR MFLD-1 (function NXOR - EQ)

Examples of other functions realizations with the OPR MFLD-1 as fragments of images are shown in Figures 11

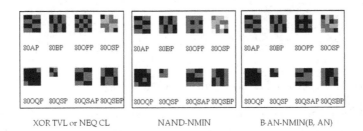

XOR TVL or NEQ CL   NAND-NMIN   B·AN-NMIN(B, AN)

Figure 11. Simulation results for other functions realizations with the OPR MFLD-1

2.5. Investigation of the base cell for the second version of OPR MFLD-2

2.5.1. Simulation of OPR MFLD-2 with OrCAD 16.3

The second circuit variant is shown in Figure 12. It differs from the previously discussed first variant that the input optical signals from each of the i,j-th base cell of two picture operands are fed to a photo-detector. One of the picture input using the appropriate shadow mask weakens the signals of one of the operands is a factor of 2. Therefore, the first unit of the circuit consists of current comparators, which convert the output voltages into a digital form that is uniquely appropriate input situation.

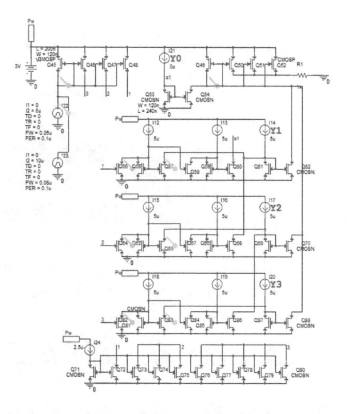

Figure 12. Circuit diagram of the base cell for the OPR MFLD-2 (the second version)

With the help of nodes in the current voltage conversion and control signals Y0-Y3 at the output node is formed by the resulting signal as a current, which corresponds to the selected desired logic function. The set of possible logical set of vector signals Y0-Y3 has 16 possible combinations. Selecting one of them allows you to implement any 16 of possible two-valued logic of binary operations. If the input signals are continuous in the time-pulse coded form, selecting the desired operation as a two-valued logic, such as AND, the operation MIN is implemented from time-pulse encoded signals. For the first model experiments in the scheme of an input photo-sensor used two of the current source to set the time of the input time-pulse signals (TPS).Instead of photo detectors are used to control the function of the sources of Y0 ÷ Y3 current. The reference currents are shown as current sources for simplicity. The current sources can be implemented on the same transistors or may be given by means of optical signals with fixed intensity. For the formation of the amplified output current which is required for light emitters, or for driver circuit, you can use the multiplier current at the current mirror.

The simulation results of this scheme on 65nm CMOS transistors with OrCAD 16.3 PSpice, at different voltages and power levels of input signals are shown in fig. 13 -20.

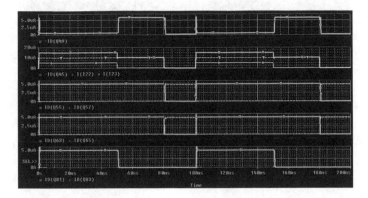

Figure 13. The results of modeling of the base cell for second version of OPR MFLD-2 for implementation of function non-equivalence of continuous logic (CL) based on XOR TVL

Experiments have shown that the power consumption of a cell does not exceed $200 \div 300 \mu W$, delay times and pulse fronts are less than 1 nanosecond, and the basic cell is realized on 44 (or 36) transistors and 11 current sources on $11 \div 15$ transistors. The duration time of pulse-coded signal is in the range of processing cycles, and the pulse period is 100 nanoseconds. This shows that it is possible to increase the frame processing rate to 10 MHz but at the expense of accuracy and complexity of matching photodetectors with current mirrors. Simulation results with OrCAD16.3 of the same basic cell circuit of the OPR MFLD-2 in the mode of implementation of the functions of the nonequivalence CL or XOR TVL are shown in Fig. 13. Diagrams that explain the work of OPR MFLD-2 in the implementation of functions of the nonequivalence CL or XOR TVL: Id = $5\mu A$, 3V supply voltage, signal durations t^a_{pulse} = 50ns, t^b_{pulse} = 80ns. In the first diagram above - the output current signal, the second - two input signals and their weighted sum, the down three: the third, fourth and fifth - currents at the output of the threshold units (green solid) and their complements (blue dashed). It uses vector tuning signals $Y= \{Y0, Y1, Y2, Y3\} = \{0, 1, 1, 0\}$, and the current level is 5 μA. At the output the correct signal is formed $\approx$ 30 ns duration. The change of the vector set to $\{0, 1, 0, 0\}$ allows for the output function I22 * NI23 (where NI23 – the complement of the signal I23), as shown in Figure 14. For credibility, that the function is implemented correctly, we did a change in the duration of signals, such that the first signal t_{pulse}(I22) = 80ns and t_{pulse} (I23) = 50ns (the signals changed their duration). The results showed that there was a signal at the output, which has a duration $\approx$30 ns.

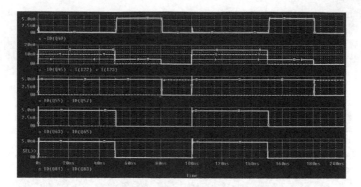

Figure 14. The diagrams of signals in the circuit with a vector set {0, 1, 0, 0} for the implementation of the function AND(a, $\bar{b}$), where a=I22, b=I23

If change of the vector set to {0, 0, 1, 0} than there is a signal at the output which differs only in the short false pulses. Change of durations of the input signals at the same vector set provides the desired signal at the output (see Figure 15). This confirms the correct operation of the scheme.

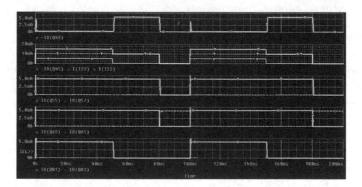

Figure 15. The diagrams of signals in the implementation of the function AND ($\bar{a}$, b)defined by the vector set {0, 0, 1, 0}, where a=I22, b=I23, t_p (I22) = 50ns, t_p(I23) = 80ns.

In Figure 16 (left) the implementation of the equivalence CL (based on NXOR TVL) is shown. The output signal (the first graph above) has the total duration of 70ns. The operation NOR TVL and on its basis the operation $\overline{max}$(a, b) CL, or the same operation min($\bar{a}$, $\bar{b}$) CL is shown in Figure 16 (right). Duration of the output signal is 20ns. Signal diagrams for mode of formation of min CL-function (based on AND) are shown in Figure 17. Left on the diagrams shows the control signals of the vector Y= {Y0, Y1, Y2, Y3} = {0, 0, 0, 1}, and the right - signals: output, input and intermediate. As can be seen from the simulations, device successfully implements the desired function when changing the supply voltage from 1,5V to 3.3V and in accordance with the results: power consumption $P_{drain} \leq 150 \mu W$ by 1.5V, current pulses amplitudes are 5µA and 10µA; power consumption $P_{drain} \leq 350 \mu W$ by 3.3V, current pulse amplitudes 5µA and 10µA.

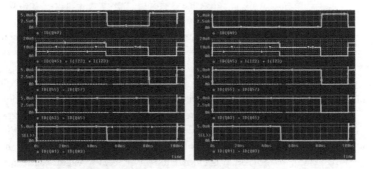

Figure 16. Simulation results of the base cell for second version of OPR MFLD-2 for implementation of function: left - equivalence operation CL (NXOR TVL), right - operation m̄ax(a, b) CL (NOR TVL)

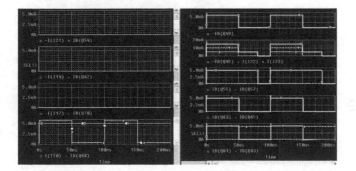

Figure 17. Signal diagrams for mode of AND (min CL) operation implementation

Circuit diagram (Figure 18) of the OPR MFLD-2 with photodiodes is used for simulations with OrCAD16.3 PSpice. The model of the photodiode is the same as in Figure 4. The simulation results are shown in Figure 19. Displaying 4 periods, at each different tuning vector set is applied and different functions is performed: the first period - vector {1,0,0,1} (equivalence), the second period - vector {1,0, 1,0} (inversion of the first variable), the third period - vector {0,1,1,0} (non-equivalence), the fourth period - vector {0,1,0,0} (AND ($\bar{a}$, b).

The signals of these vectors are displayed on the lower four graphs yellow lines. The blue lines show the output currents generated configuration signals and the corresponding nodes. The sum of output currents of these nodes represents the output signal. It was featured on the second chart above the green line and the input photocurrent from the two arguments shows a blue line. At the top graph shows the power consumption of the base cell. The main problem in these cells is a significant deterioration in fronts (an increase of up to 200 ns). Moreover, no change in the operating voltage from 3V to 5V, no change in amplitude of photocurrents (in the experiments, Io = 5µA, 10µA, 15µA, but at 20µA did not work!), including at different levels of reference current generators, practice does not significantly affect the duration of the fronts. It is therefore necessary to look for other circuit solutions, for example, use the cascode circuit of current mirrors, more complex, but high-speed, current or voltage comparators. But at the same time significantly increase the hardware cost of a basic cell, and it does not allow for a high level of integration on a chip. So here we are showing the circuit with extended processing period up to 10µS, which with Io = 5µA circuit will provide the required characteristics. Power consumption does not exceed 300÷350µW at a supply voltage of 2.4V and the 3.0V on photodiodes. Results of experiments are shown in Figure 20. By dynamic reconfiguration of optical signals (vector Y) the desired function of the basic cell is provided and duration of the reconfiguration process is equal to the period T = 10÷100µs. In addition, if use other technologies, the vectors set can be represented using electrical signals.

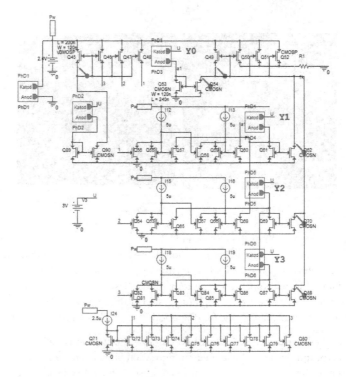

Figure 18. The base cell for OPR MFLD-2 with one input and four control photodiodes

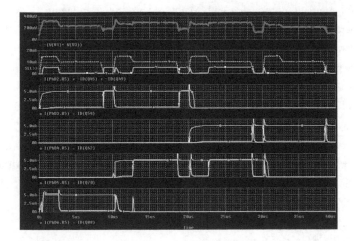

Figure 19. Diagrams of signals at modeling cell with an optical configuration for the desired function and the input photodiode

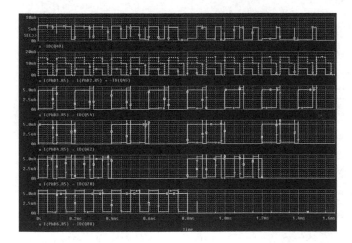

Figure 20. Diagrams showing the ability to dynamically reconfigurable the cells on the implementation of all 16 possible functions of TVL with period of 100μs (total duration 16 periods). The first graph shows the output signal and the second - the input signals. At the bottom four graphs in yellow show signals at photodiodes, and green - generated current logical components

2.5.2. Simulation of the OPR MFLD-2 with MathCAD

Simulation results of the offered OPR MFLD-2 with MathCAD and it usage for image processing and fuzzy logic operations are shown in fig. 21-24.

Formulas for simulation processing with MathCAD are shown in Figure 21. At first, input two 2D operands $\Lambda 1$ and **B1** and its weighted sum **SIAB** are formed. The coefficient and threshold $t_0 = 10$ because the current in the OPR MFLD-2 circuit is 10µA. Contrast complementary images are matrixes **AN1** and **BN1**. After threshold processing by current comparators the direct matrixes **T1SIAB, T2SIAB, T3SIAB** and matrixes **TN1SIAB, TN2SIAB, TN3SIAB** of complementary images are formed. Four picture tuning operand **NY0 ÷NY3** are formed with tuning vector signals ny0÷ny3. Four logical members **SY0÷SY3** are formed using simultaneous threshold and state decoding operations. The sum of those members is the output matrix function **NF**. All operands dimension is 64x64 elements. All images of above mentioned matrixes and some output functions are shown in Figure 22.

$D1 := \text{READBMP}(\text{"D:\TatoD\tato2\tato\ff1.bmp"})$

$D2 := \text{READBMP}(\text{"D:\TatoD\tato\tato\ff2.bmp"})$

D1

$A1 := \text{submatrix}\left[\left(D1 \cdot \dfrac{10}{255}\right), 0, 63, 0, 63\right] \qquad AN1_{i,j} := 10 - A1_{i,j}$

$B1 := \text{submatrix}\left[\left(D2 \cdot \dfrac{10}{255}\right), 0, 63, 0, 63\right] \qquad BN1_{i,j} := 10 - B1_{i,j}$

$i := 0..63 \quad j := 0..63 \qquad R_{i,j} := 1 \quad RN_{i,j} := 0$

$tr := 30 \qquad to := 10$

$XD_{i,j} := 10 \cdot \Phi(65 - i - j)$

$YD_{i,j} := 10 \cdot \Phi[(63 - j \cdot 2) + i - 60]$

$SIAB := A1 + 2 \cdot B1$

$ny0 := 0 \quad ny1 := 1 \quad ny2 := 1 \quad ny3 := 1$

$T1SIAB_{i,j} := 10 \cdot \Phi\left(SIAB_{i,j} - 10\right) \qquad NY0_{i,j} := 10 \cdot ny0 \quad NY1_{i,j} := 10 \cdot ny1$

$T2SIAB_{i,j} := 10 \cdot \Phi\left(SIAB_{i,j} - 20\right) \qquad NY2_{i,j} := 10 \cdot ny2 \quad NY3_{i,j} := 10 \cdot ny3$

$T3SIAB_{i,j} := 10 \cdot \Phi\left(SIAB_{i,j} - 30\right)$

$TN1SIAB_{i,j} := 10 - T1SIAB_{i,j}$

$TN2SIAB_{i,j} := 10 - T2SIAB_{i,j}$

$TN3SIAB_{i,j} := 10 - T3SIAB_{i,j}$

$SY0_{i,j} := \Phi\left(NY0_{i,j} - T1SIAB_{i,j}\right) \cdot \left(NY0_{i,j} - T1SIAB_{i,j}\right)$

$SY1_{i,j} := \Phi\left(NY1_{i,j} - T2SIAB_{i,j} - TN1SIAB_{i,j}\right) \cdot \left(NY1_{i,j} - T2SIAB_{i,j} - TN1SIAB_{i,j}\right)$

$SY2_{i,j} := \Phi\left(NY2_{i,j} - T3SIAB_{i,j} - TN2SIAB_{i,j}\right) \cdot \left(NY2_{i,j} - T3SIAB_{i,j} - TN2SIAB_{i,j}\right)$

$SY3_{i,j} := \Phi\left(NY3_{i,j} - TN3SIAB_{i,j}\right) \cdot \left(NY3_{i,j} - TN3SIAB_{i,j}\right)$

$NF_{i,j} := SY0_{i,j} + SY1_{i,j} + SY2_{i,j} + SY3_{i,j}$

Figure 21. Formulas for simulation of OPR MFLD-2 with MathCAD

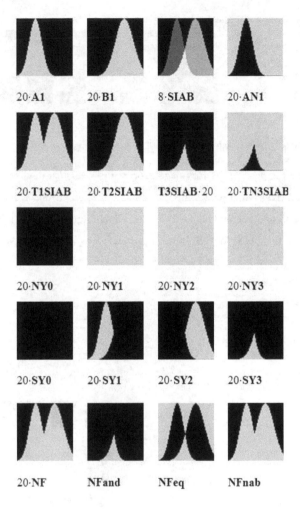

Figure 22. The simulation results of the OPR MFLD-2 with MathCAD for single-cycle high-speed computation of continuous logic operations and / or fuzzy logic for membership functions. In the bottom row the functions realization MAX/OR, MIN/AND, EQ/NXOR, $(\bar{A} \cdot B)$ over the two graphs presented in the form of membership functions of operands **A1** and **B1**

Simulation results for different functions (AND, EQ, NEQ, OR) implementation in four different sub-regions is shown in Figure 23. **XD** and **YD** are the input matrixes. Tuning matrixes **VY0÷VY3** have different values in sub-regions. Output matrix **VF** is concatenation of sub-region functions.

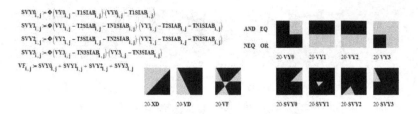

Figure 23. The simulation results for sub-region function AND, EQ, NEQ, OR implementation

Let's demonstrate the possibilities for image processing with such devices. An example of contour extraction (**NF**) when processing the first input operand image A1 and its shifted copy **AES1** as the second operand is shown in figure 24. In figure 24: **NY0, NY1, NY2, NY3** – tuning matrixes for that operation; **NF** – the output image.

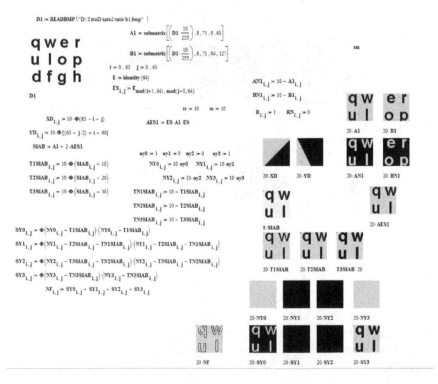

Figure 24. Simulation results of the OPR MFLD-2 with MathCAD for contour extraction

3. Conclusions

We have developed two version of OPR MFLD which realizes the universal binary logic on optical signals. They have subpixel configuration of 2x2 elements, consist of a small amount of photodiodes (4) and transistors (18), have low power consumption <1-5mW, high productivity and realize the basic set of operations of continuous logic with time pulse representation of processed signals. Modeling of such cells with OrCad is made. It is confirmed that all set of possible functions will be realized with such MFLD by a simple photo tuning. Such cells for OPR MFLD are integrated into array of 32x32 allow reaching productivity 10^{12} CL-logic operations/sec.

Author details

Vladimir G. Krasilenko[1*], Aleksandr I. Nikolskyy[2*] and Alexander A. Lazarev[2]

*Address all correspondence to: krasilenko@mail.ru,fortuna888@i.ua

1 Vinnitsa Social Economy Institute of Open International University of Human Development "Ukraine", Ukraine

2 Vinnitsa National Technical University, Ukraine

References

[1] Semiconductor Industry Association. (2001). The international Technology Roadmap for Semiconductors San Jose, USA.

[2] Fey, D. (2001). Architecture and technologies for an optoelectronic VLSI. *Optik*, 112(7), 274-282.

[3] Li, G, Huang, D, Yucetukk, E, Marchand, P, Esener, J, Ozguz, S, & Liu, Y. (2002). There- Dimensional Optoelectronic Stacked Processor by use of Free-Spase Optical Interconnection and Three-Dimensional VLSI Chip Stacks. *Applied optics.*, 41.

[4] Honeywell Technology Center. http://htchoneywell.com/photonics.

[5] Masahiko, M, & Toyohiko, Y. (1997). Optical learning neural networks with two dimensional structures. *Proc. SPIE.*, 3402, 226-232.

[6] Berger, C, Collings, N, & Gehriger, D. (1997). Recurrent Optical Neural Network for the Study of Pattern Dynamics. *Proc. SPIE.*, 3402, 233-244.

[7] Krasilenko, V. G, Nikolskyy, A. I, Zaitsev, A. V, & Voloshin, V. M. (2001). Optical pattern recognition algorithms on neural-logic equivalent models and demonstration of their prospects and possible implementations. *Proc. SPIE,,* 4387, 247-260.

[8] Krasilenko, V. G, Saletsky, F. M, Yatskovsky, V. I, & Konate, K. (1997). Continuos logic equivalental models of Hamming neural network architectures with adaptive correlated weighting. *Proc. SPIE,* 3402, 398-408.

[9] Krasilenko, V. G, Nikolskyy, A. I, Lazarev, A. A, & Sholohov, V. I. (2004). A concept of biologically motivated time-pulse information processing for design and construction of multifunctional devices of neural logic. *Proc. SPIE,* 5421, 183-195.

[10] Krasilenko, V. G. (1988). Optoelectronic structures in information-measuring systems for image processing. Dissertation for the degree of candidat of tech. science. Vinnitsa., 188.

[11] Krasilenko, V. G, Kolesnitsky, O. K, & Boguhvalsky, A. K. (1995). Creation Opportunities of Optoelectronic Continuous Logic Neural Elements, which are Universal Circuitry Macrobasis of Optical Neural Networks. *Proc. SPIE,* 2647, 208-217.

[12] Huang, H, Itoh, M, & Yatagai, T. (1999). Optical scalable parallel modified signed-digit algorithms for large-scale array addition and multiplication using digit-decomposition-plane representation. *Opt. Eng.,* 38, 432-440.

[13] Huang, K. S, Yenkins, B. K, & Sawchuk, A. A. (1989). Image algebra representation of parallel optical binary arithmetic. *Applied Optics,* 28, N6, 1263-1278.

[14] Awwal, A. A, & Iftekharuddin, K. M. (1999). Computer arithmetic for optical computing. *Special Section Opt. Eng.,* 38.

[15] Krasilenko, V. G, Yatskovsky, V. I, & Dubchak, V. N. (2001). Organization and design of computing structures of matrix-quaternion sign-digit arithmetic. *Measuring and Computer Technique in Technological Processes ,* 1, 146-150.

[16] Guilfoyle, P. S, & Mccallum, D. S. (1996). High-speed low-energy digital optical processors. *Opt. Eng.,* 35(2), 436-442.

[17] Krasilenko, V. G, & Dubchak, V. N. (1988). Structure of optoelectronic processors of parallel type for image processing. *New electronic installations and devices,,* 63-65.

[18] Krasilenko, V. G, Kolesnitsky, O. K, & Dubchak, V. N. (1991). Creation principles and circuitry construction question of multichannel arrangements and systems for parallel image analysis and processing. *Proc. 1-st All-Union Meeting on Pattern recognition and image analysis,* Minsk , 3, 83-87.

[19] Krasilenko, V. G, & Magas, A. T. (1999). Fundamentals of design of multi-functional devices of matrix multi-valued logic with fast programmed adjusting. *Measuring and computer technique in technological processes,* 4, 113-121.

[20] Krasilenko, V. G, Dubchak, V. N, & Boyko, R. (1990). Development and application of optoelectronic 2D-array bi-stable structures. *Information Bulletin of Belarus Academy of Sciences*, 6, 69-72.

[21] Krasilenko, V. G, Nikolsky, A. I, Yatskovsky, V. I, Ogorodnik, K. V, & Lischenko, S. (2002). The family of new operations "equivalency" of neuro-fuzzy logic, their optoelectronic realization and applications. *Proc. SPIE*, 4732, 106 -120 .

[22] Shimbirev, P. N. (1990). Hybrid continuous-logic arrangements. Moscow: Energoizdat., 176.

[23] Levin, V. I. (1990). Continuous logic, its generalizationand application. *Automatica and telemechanica,*, 8, 3-22.

[24] Samofalov, K. G, Korneychuk, V. N, Romankevich, A. I, & Tarasenko, V. P. (1974). Digital multi-valued logic devices. Kiev: Higher school, 168.

[25] Krasilenko, V. G, Nikolskyy, A. I, Lazarev, A. A, & Pavlov, S. N. (2005). Design and applications of a family of optoelectronic photocurrent logical elements on the basis of current mirror and comparators. *Proc. SPIE*, 5948, 426-435.

[26] Krasilenko, V. G, Nikolskyy, A. I, & Lazarev, A. A. (2009). Perfection of circuits for realization of the generalized equivalence (nonequivalence) operations neurobiologic. *Announcer of the Khmelnytsk national university.*, 2, 174-178.

[27] Krasilenko, V. G, Nikolskyy, A. I, Lazarev, A. A, & Lobodzinska, R. F. (2007). Multithreshold comparators with synhronous control of thresholds. *Proc. of «Dynamika naykowych badan- 2007"*, 8, Technical science: Przemysl. Nauka i studia:, 55-58.

Chapter 3

Optoelectronic Oscillators Phase Noise and Stability Measurements

Patrice Salzenstein

Additional information is available at the end of the chapter

1. Introduction

A classical method to characterize the spectral density of phase noise of microwave oscillators is to compare the device under test (DUT) to another one, we call it the reference, with the same frequency if noise is expected to be better for the reference. In most of case it is possible to then characterize oscillators or synthesizers. But it is only possible if we can assign the same frequency to the DUT and the reference signal. However for some applications, we see that the DUT delivers a frequency hard to predict during the fabrication. We here focus on characterizing a special class of oscillators, called optoelectronic oscillators (OEO) [1]. An OEO is generally an oscillator based on an optical delay line and delivering a microwave signal [2]. Purity of microwave signal is achieved thanks to a delay line inserted into the loop. For example, a 4 km delay corresponds to a 20 µs-long storage of the optical energy in the line. The continuous optical energy coming from a laser is converted to the microwave signal. This kind of oscillators were investigated [3]. By the way, such OEO based on delay line are still sensitive to the temperature due to the use of optical fiber. Recently, progress were made to set compact OEO thanks to optical mini-resonators or spheres [4 - 7]. Replacing the optical delay line with an ultra-high Q whispering gallery-mode optical resonator allows for a more compact setup and an easier temperature stabilization. In order to introduce into the loop the fabricated resonator in MgF_2 [8], CaF_2 or fused silica, it has to be coupled to the optical light coming from a fiber. Best way to couple is certainly to use a cut optical fiber through a prism. But a good reproducible way in a laboratory is to use a tapered fibber glued on a holder. Then appears a problem for determining the phase noise of such compact OEO, because the frequency is rarely predictable. It is impossible to choose the frequency of the modulation (for instance exactly 10 GHz) and that's why it is necessary to develop new instruments and systems to determine phase noise for any delivered signal in X-band (8.2-12.4 GHz) for instance. Here we come to our goal. The aim of this chapter is to provide a tool for a better knowledge of the phase noise characterization of OEO's using an optoelectronic phase noise system. We logically start by giving the main principle of how work such a system. We see that the main idea is to use delay lines to perform phase noise measurements. We also consderably increase the performances of such a system by cross correlation measurement, thanks to the two quasi-identical arms developed instrument. Then we present a realized system and the evaluation of its uncertainty based on international standard for determination of uncertainties when characterizing an OEO in X-band.

2. Principle of the phase noise measurement system

A quasi-perfect RF-microwave sinusoidal signal can be written as :

$$v(t) = V_0[1 + \alpha(t)]cos(2\pi v_0 t + \varphi(t)) \tag{1}$$

where V_0 is the amplitude, v_0 is the frequency, $\alpha(t)$ is the fractional amplitude fluctuation, and $\varphi(t)$ is the phase fluctuation. Equation (1) defines $\alpha(t)$ and $\varphi(t)$ in low noise conditions: $|\alpha(t)| \ll 1$ and $|\varphi(t)| \ll 1$. Short-term instabilities of signal are usually characterized in terms of the single sideband noise spectral density PSD S(f). Phase noise £(f) is typically expressed in units of dBc/Hz, representing the noise power relative to the carrier contained in a 1 Hz bandwidth centered at a certain offsets from the carrier. So, S is typically expressed in units of decibels below the carrier per hertz (dBc/Hz) and is defined as the ratio between the one-side-band noise power in 1 Hz bandwidth and the carrier power:

$$£(f) = \frac{1}{2} \cdot S\varphi(f) \tag{2}$$

This definition given in equation (2) includes the effect of both amplitude and phase fluctuations. Phase noise is the frequency domain representation of rapid, short-term, random fluctuations in the phase of a waveform, caused by time domain instabilities. However we must know the amplitude and phase noise separately because they act differently in the circuit. For example, the effect of amplitude noise can be reduced by amplitude limiting mechanism and mainly suppressed by using a saturated amplifier. Phase noise of microwave oscillators can usually be characterized by heterodyne measurement. Whereas, for such a system, we need a reference oscillator operating exactly at the frequency of the DUT with lower phase noise. Phase noise can be measured using a spectrum analyzer if the phase noise of the device under test (DUT) is large with respect to the spectrum analyzer's local oscillator. Care should be taken that observed values are due to the measured signal and not the Shape Factor of the spectrum analyzer's filters. Spectrum analyzer based measurement can show the phase-noise power over many decades of frequency. The slope with offset frequency in various offset frequency regions can provide clues as to the source of the noise, e.g. low frequency flicker noise decreasing at 30 dB per decade.

Reference is no more required for homodyne measurement with a delay line discriminator. At microwave frequencies, electrical delay is not suitable because of its high losses. However photonic delay line offers high delay and low attenuation equal to 0.2 dB/km at the wavelength λ=1.55 μm. Optoelectronic phase noise measurement system is schematically represented on Figure 1.

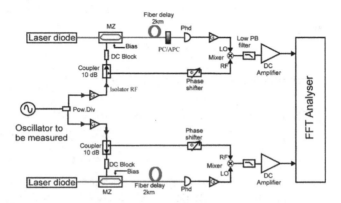

Figure 1. Phase noise bench.

It consists on two equal and fully independent channels. The phase noise of the oscillator is determined by comparing phase of the transmitted signal to a delayed replica through optical delay using a mixer. It converts the phase fluctuations into voltage fluctuations. An electro-optic modulator allows modulation of the optical carrier at microwave frequency. The length of the short branch where microwave signal is propagating is negligible compared to the optical delay line. Mixers are used as phase detectors with both saturate inputs in order to reduce the amplitude noise contribution. The low pass filters are used to eliminate high frequency contribution of the mixer output signal. DC amplifiers are low flicker noise.

The oscillator frequency fluctuation is converted to phase frequency fluctuation through the delay line. If the mixer voltage gain coefficient is K_φ (volts/radian), then mixer output rms voltage can be expressed as :

$$V^2_{out}(f) = K^2_\varphi \mid H_\varphi(jf) \mid ^2 S_\varphi(f) \tag{3}$$

Where $\mid H_\varphi(jf) \mid^2 = 4.\sin^2(\pi f \tau)$ is the transfer function of optical delay line, and f is the offset frequency from the microwave carrier. Equation (3) shows that the sensitivity of the bench depends directly on K^2_φ and $\mid H_\varphi(jf) \mid$. The first is related to the mixer and the second essentially depends on the delay τ. In practice, we need an FFT analyzer to measure the spectral density of noise amplitude $V^2_{out}(f)/B$, where B is the bandwidth used to calculate $V_{out}(f)/B$. The phase noise of the DUT is finally defined by Eq. (4) and taking into account the gain of DC amplifier G_{DC} as :

$$\pounds(f)=[V^2_{out}(f)]\cdot/\cdot[2K^2_\varphi\cdot\mid H_\varphi(jf)\mid{}^2G^2_{DC}\cdot B]\tag{4}$$

Such instruments has been recently introduced [3,9,10]. In section 3, we present concretely a realized optoelectronic phase noise measurement system.

3. Description of the realized system

In this section we concretely present a realization. We apply the principle detailed in the previous section to settle a phase noise optoelectronic system. For the demonstration, we characterize a frequency synthesizer as a DUT. It presents advantage to check different frequencies in X-band. System is shown on Figure 2.

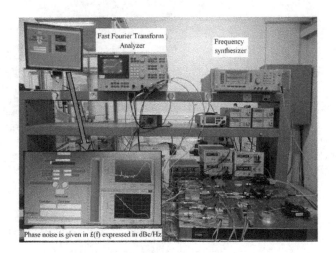

Figure 2. Picture of the phase noise measurement system.

The system is composed from different parts. We see on Figure 2, that we use a frequency synthesizer to check if the system works properly in X-band. On the picture we see the results of the phase noise characterization (inserted on the left of the picture) for a +3 dBm, 10 GHz signal. On the top of the picture we see the double channels Fast Fourier Transform analyzer used for this purpose (*Hewlett-Packard* HP3561A). £(f) expressed in dBc/Hz is deduced from the data provided by the FFT analyzer are given in V²/Hz.

Figure 3 shows the picture of the *Hewlett-Packard* HP3561A FFT analyzer. Note that the data are expressed in V²/Hz. It is necessary to use a program to get the expected quantity £(f) in dBc/Hz. It is developed in the next section of this chapter.

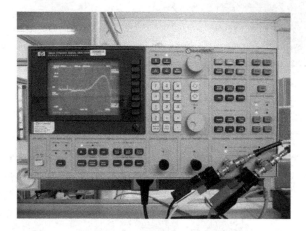

Figure 3. Picture of the double channels *Hewlett-Packard* HP3561A FFT analyzer.

4. Validation of the performances

The measured phase noise includes the DUT noise and the instrument background. The cross correlation method allows to decrease the cross spectrum terms of uncommon phase noise as $\sqrt{(1/m)}$, where m is the average number. Thereby uncorrelated noise is removed and sensitivity of measure is improved. To validate the measure of our phase noise bench, we need to compare data sheet of the commercial frequency synthesizer Anritsu/Wiltron 69000B [11] with the phase noise we measure using our system.

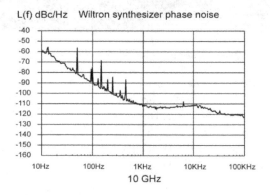

Figure 4. Phase noise (dBc/Hz) of the synthesizer measured at 10 GHz with K_φ=425 mV/rad and G_{DC}= 40dB versus Fourier Frequency between 10 Hz and 100 kHz.

Figure 4 shows the result of this measure. We can see that our bandwidth is limited to 100 kHz (τ = 10 μs) and the measured phase noise corresponds to the data sheet.

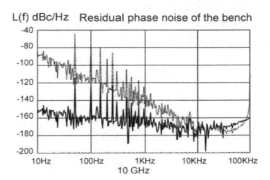

L(f) dBc/Hz   Residual phase noise of the bench

Figure 5. Phase noise floor (dBc/Hz) of the bench measured at 10 GHz with Anritsu synthesizer (500 averages) versus Fourier Frequency between 10 Hz and 100 kHz.

Figure 5 represents the background phase noise of the bench after performing 500 averaged with cross-correlation method, when removing the 2 km optical delay line. In this case, phase noise of the 10 GHz synthesizer is rejected. The solid curve shows noise floor (without optical transfer function) respectively better than -150 and -170 dBc/Hz at 10^1 and 10^4 Hz from the 10 GHz carrier. Dotted curve is the noise floor when optical fiber is introduced.

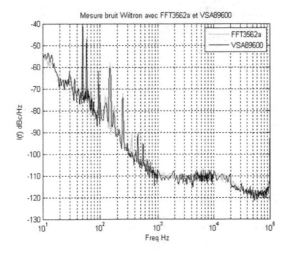

Figure 6. Spectral density of phase noise floor £(f) expressed in dBc/Hz versus Fourier frequencies (in Hz) between 10 Hz and 100 kHz for a commercial synthesizer measured with our bench with Kφ = 425 mV/rad and GDC = 40 dB for two different FFT analyzing system : *Hewlett-Packard* 3562A and *Agilent* 89600.

One can note that the use of a shorter delay line in a optoelectronic phase noise measurement system working in X-band, allow a characterization of the phase noise far from the carrier. Fourier frequency analysis can be extended from 10^5 to 2×10^6 Hz by introducing a 100 m delay line in addition of a 2 km optical fiber [12]. As the Fast Fourier Transform (FFT) analyzer (*Hewlett-Packard* 3562A) used for characterizing the noise up to 100 kHz from the carrier is not operating for higher frequencies, it is necessary to use another FFT such as an *Agilent* 89600 for instance.

On Figure 6, we check that the two different FFT systems provide the same results for Fourier frequencies between 10 Hz and 100 kHz.

Note that our results are as expected with the data sheet of a 10 GHz phase noise spectrum for an *Wiltron* 69000B series. Our bandwidth is limited to 100 kHz ($\tau = 10$ μs) and the measured phase noise corresponds to the data sheet. Figure 6 then gives the noise floor of the instrument versus Fourier frequencies. The noise floor is respectively better than – 90 and – 170 dBc/Hz at 10^3 and 5×10^6 Hz from the carrier.

The introduction of a 100 m short fiber in addition to the 2 km fiber allows to characterize phase noise of oscillators with an optoelectronic phase noise measurement system.

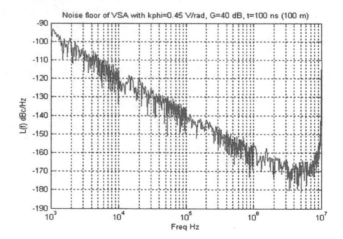

Figure 7. Spectral density of phase noise floor £(f) expressed in dBc/Hz versus Fourier frequencies (in Hz) for the developed system when using the 100 meters delay lines.

100 m fiber corresponds to a 500 ns delay. So the 10^5 Hz limit due to the 2 km fiber can be extended from 10^5 to 2×10^6 Hz from the carrier. This system works for any microwave signal in X-band (8.2 – 12.4 GHz) especially for those delivered by optoelectronic oscillators. We see on Figure 7 that the noise floor of the system is in the range of –165 dBc/Hz at 5. 10^6 Hz from the X-band microwave signal.

5. Evaluation of the uncertainty of such a measurement system

This evaluation is based on a previous work [13]. For the evaluation of the uncertainty, we use the main guideline delivered by the institution in charge of international metrology rules, *Bureau International des Poids et Mesures* (BIPM) in the guide "Evaluation of measurement data — Guide to the expression of uncertainty in measurement" [14]. According to this guideline, the uncertainty in the result of a measurement generally consists of several components which may be grouped into two categories according to the way in which their numerical value is estimated. The first category is called "type A". It is evaluated by statistical methods such as reproducibility, repeatability, special consideration about Fast Fourier Transform analysis, and the experimental standard deviation. The components in category A are characterized by the estimated variances. Second family of uncertainties contributions is for those which are evaluated by other mean. They are called "type B" and due to various components and temperature control. Experience with or general knowledge of the behavior and properties of relevant materials and instruments, manufacturer's specifications, data provided in calibration and other certificates (noted BR), uncertainties assigned to reference data taken from handbooks. The components in category B should be characterized by quantities which may be considered as approximations to the corresponding variances, the existence of which is assumed.

5.1. "Type A". Statistical contributions

Uncertainty on $£(f)$ strongly depends on propagation of uncertainties through the transfer function as deduced from equation (3). We see here that "type A" is the main contribution. For statistical contributions aspects, the global uncertainty is strongly related to repeatability of the measurements. Repeatability (noted A1) is the variation in measurements obtained by one person on the same item and under the same conditions. Repeatability conditions include: the same measurement procedure, the same observer, the same measuring instrument, used under the same conditions, repetition over a short period of time, the same location. We switch on all the components of the instrument and perform a measurement keeping the data of the curve. Then we need to switch them off and switch them on again to obtain another curve. We must repeat this action several times until we have ten curves. Elementary term of uncertainty for repeatability is experimentally found to be equal to 0.68 dB.

Other elementary terms of statistical contributions still have lower contributions.

Reproducibility, noted A2, is the variability of the measurement system caused by differences in operator behavior. Mathematically, it is the variability of the average values obtained by several operators while measuring the same item. The variability of the individual operators are the same, but because each operator has a different bias, the total variability of the measurement system is higher when three operators are used than when one operator is used. When the instrument is connected, there is no change of each component or device inside. That makes this term negligible because, for example, the amplifier is never replaced by another one. We remark that, if one or more of the components would be replaced, it will be necessary to evaluate the influence of the new component on the results of measurements. Main aspect is that operators are the same. In our case, operator don't change. It means a first approximation we can take 0 dB for this uncertainty.

Uncertainty term due to the number of samples is noted A3. It depend on how many points are chosen for each decades. For this term, we check how works the Fast Fourier Transform (FFT) analyzer, it leads to an elementary term of uncertainty less than 0.1 dB.

Finally, statistical contribution can be considered as:

$$A = \sqrt{(\sum A_i{}^2)} \tag{5}$$

According to equation (5), it can then be considered that the whole statistical contribution is better than 0.69 dB.

5.2. "Type B". Other means

Our system is not yet formally traceable to any standard, according to how the phase noise is determined. Phase noise measurement generally don't need to be referenced to a standard as the method is intrinsic. So the data provided in calibration and other certificates, noted BR, are not applicable. It results that we can take 0 dB as a good approximation of BR.

Influence of the gain of the DC amplifier, noted BL1, is determined to be less than 0.04 dB.

Temperature effects, noted BL2, are less than 0.1 dB as optical fiber regulation system of temperature in turned on.

Resolution of instruments, noted BL3, is determined by the value read on each voltmeter when we need to search minimum and maximum for the modulator but also for wattmeter. Resolution is then no worse than 0.1 dB.

During the measurement, what influence could bring the DUT, generally an oscillator to be characterized? We call BL4 the influence of the variation of the DUT. It stays negligible while the variation stay limited. It results that we can keep 0 dB as a reasonable value for BL4 if the DUT is not too unstable.

Uncertainty on the determination of the coefficient K_φ (noted BL5) dependent for the slope expressed in Volt/rad is found to be lower than 0.08dB.

For the contribution of the use of automatic/manual range, noted BL6, we can deduce from experimental curves that this influence is no more than 0.02 dB. In our case, all these terms were found lower than repeatability.

The influence of the chosen input power value of the DUT, noted BL7, is negligible as the input power in normal range i. e. between -4 dBm and +6 dBm, has a very limited influence. We experimentally observe an influence. It stay better than 0.02 dB.

Other elementary terms still have lower contributions. BL is the arithmetic sum of each elementary contribution. It is determine to be 0.38 dB.

5.3. Estimation of the global uncertainty of this system

According to the Guide to the expression of uncertainty in measurement, uncertainty at 1 sigma interval of confidence is calculated as follows:

$$u_c = \sqrt{(A^2 + BR^2 + BL^2)} \tag{6}$$

We deduce from equation (6) that uncertainty at 1 sigma, noted u_c, is better than 0.79 dB.

Table 1 summarizes how is deduced the global uncertainty for the spectral density of phase noise at 1 sigma.

Uncertainty	Designation	Value (in dB)
A1	Repeatability	0.68
A2	Reproductibility	0
A3	Uncertainty term due to the number of sample	0.1
A	$(\Sigma Ai^2)^{1/2}$	0.69
BR	Not applicable	0
BL1	Gain of the DC amplifier	0.04
BL2	Influence of the temperature	0.1
BL3	Influence of the resolution of the instrument	0.1
BL4	Influence of the power of the DUT	0
BL5	Uncertainty on the determination of K_ϕ	0.08
BL6	Contribution of automatic/manual range	0.02
BL7	Influence of the variation of the input power	0.02
BL	ΣBLi	0.38
u_c	Global uncertainty at 1 sigma: $(A^2+BR^2+BL^2)^{1/2}$	0.79

Table 1. Budget of uncertainties.

Its leads to a global uncertainty of $U = 2xu_c = 1.58$ dB at 2 sigma.

For convenience and to have an operational uncertainty in case of degradation or drift of any elementary term of uncertainty, it is wise to degrade the global uncertainty. That's why we choose to keep $U = 2$ dB at 2 sigma for a common use of the optoelectronic system for phase noise determination. According to this evaluation of the uncertainty at 1 sigma, it leads to 1.58 at two sigma. It concretely means that it is possible to determine the phase noise of a single oscillator in X-band with a global uncertainty set to be better than ±2 dB.

6. Conclusion

The author wish that the phase noise optoelectronic system presented in this chapter is useful for those who want to understand how the phase noise can be experimentally determined. We detailed performances and consideration about estimation of the uncertainty to show the main advantage of such developed instrument for metrology or telecommunication applications and characterizations of compact OEO's operating in X-band. With high performance better than -170 dBc/Hz at 10 kHz from the 10 GHz carrier, it is interesting to underline that it is possible to determine the phase noise of a single oscillator in X-band with a global uncertainty set to be better than ±2 dB. This system is to be extended at lower and higher microwave operating frequencies.

Acknowledgements

Author would like to thank his colleagues from FEMTO-ST, Besançon, Ms. Nathalie Cholley, Engineer, Dr. Abdelhamid Hmima and Dr. Yanne K. Chembo, CNRS senior researcher, for fruitful discussions. Author also acknowledges the French National Research Agency (ANR) grant "ANR 2010 BLAN 0312 02".

Author details

Patrice Salzenstein*

Address all correspondence to: patrice.salzenstein@femto-st.fr

Centre National de la Recherche Scientifique (CNRS), Unité Mixte de Recherche (UMR), Franche Comté Electronique Mécanique Thermique Optique Sciences et Technologies (FEMTO-ST),, France

References

[1] Salzenstein, P. (2011). Optoelectronic Oscillators. In: Sergiyenko O. (ed.) Optoelectronic Devices and Properties. Rijeka: InTech;. . Available from http://www.intechopen.com/books/optoelectronic-devices-and-properties/optoelectronic-oscillators (accessed 5 June 2012)., 401-410.

[2] Yao, X. S., & Maleki, L. (1994). High frequency optical subcarrier generator. *Electronics Letters*, 30(18), 1525-1526.

[3] Volyanskiy, K., Cussey, J., Tavernier, H., Salzenstein, P., Sauvage, G., Larger, L., & Rubiola, E. (2008). Applications of the optical fiber to the generation and to the meas-

urement of low-phase-noise microwave signals. *Journal of the Optical Society of America B*, 25(12), 2140-2150.

[4] Ilchenko, V. S., Yao, X. S., & Maleki, L. (1999). High-Q microsphere cavity for laser stabilization and optoelectronic microwave oscillator. *Proceedings of SPIE*, 3611-190.

[5] Volyanskiy, K., Salzenstein, P., Tavernier, H., Pogurmirskiy, M., Chembo, Y. K., & Larger, L. (2010). Compact Optoelectronic Microwave Oscillators using Ultra-High Q Whispering Gallery Mode Disk-Resonators and Phase Modulation. *Optics Express .*, 18(21), 22358-22363.

[6] Schliesser, A., & Kippenberg, T. J. (2010). Cavity Optomechanics with Whispering-Gallery Mode Optical Micro-Resonators. *Acta Avances in Atomic Molecular and Optical Physica*, 58-207.

[7] Salzenstein, P., Tavernier, H., Volyanskiy, K., Kim, N. N. T., Larger, L., & Rubiola, E. (2009). Optical Mini-disk resonator integrated into a compact optoelectronic oscillator. *Acta Physica Polonica A*, 116(4), 661-663.

[8] Tavernier, H., Salzenstein, P., Volyanskiy, K., Chembo, Y. K., & Larger, L. (2010). Magnesium Fluoride Whispering Gallery Mode Disk-Resonators for Microwave Photonics Applications. *IEEE Photonics Technology Letters*, 22(22), 1629-1631.

[9] Salzenstein, P., Cussey, J., Jouvenceau, X., Tavernier, H., Larger, L., Rubiola, E., & Sauvage, G. (2008). Realization of a Phase Noise Measurement Bench Using Cross Correlation and Double Optical Delay Line. *Acta Physica Polonica A*, 112(5), 1107-1111.

[10] Salzenstein, P., Cholley, N., Zarubin, M., Pavlyuchenko, E., Hmima, A., Chembo, Y. K., & Larger, L. (2011). Optoelectronic phase noise system designed for microwaves photonics sources measurements in metrology application. Proceedings of SPIE ., 8071-807111.

[11] Anritsu (2000). Typical datasheet Anritsu 69B serie available on page at the following link: http://cem.inrets.fr/private/materiel-labo/images/m_011_doc_gene_65ghz.pdf (accessed 5 June 2012)., 8.

[12] Salzenstein, P., Hmima, A., Zarubin, M., Pavlyuchenko, E., & Cholley, N. (2012). Optoectronic phase noise measurement system with wideband analysis. *Proceedings of SPIE*, 8439, 84391M.

[13] Salzenstein, P., Pavlyuchenko, E., Hmima, A., Cholley, N., Zarubin, M., Galliou, S., Chembo, Y. K., & Larger, L. (2012). Estimation of the uncertainty for a phase noise optoelectronic metrology system. *Physica Scripta* [T149], 014025, http://dx.doi.org/ 10.1088/00318949 /2012/T149/014025.

[14] GUM: (2008). Guide to the Expression of Uncertainty in Measurement, fundamental reference document. , JCGM 100: (GUM 1995 minor corrections) http:// www.bipm.org/en/publications/guides/gum.html (accessed 5 June 2012).

Chapter 4

All-Optical Autonomous First-in–First-out Buffer Managed with Carrier Sensing of Output Packets

Hiroki Kishikawa, Hirotaka Umegae,
Yoshitomo Shiramizu, Jiro Oda, Nobuo Goto and
Shin-ichiro Yanagiya

Additional information is available at the end of the chapter

1. Introduction

Along with rapid progress of optical fiber links in the physical layer of networks, optical processing in the control layer such as data links and internet layers is expected to realize photonic networks. Various kinds of architectures of optical routers and switches have been exploited. Optical buffering is one of the indispensable key technologies for avoiding packet collision in these network nodes.

Various optical buffering systems have been reported [1,2]. Most of them consist of optical fiber delay lines (FDLs). Although optical slow light can be a potential candidate to adjust short delay timing [3,4,5], FDLs are regarded to be most useful elements for packet buffering. Basically, two kinds of architectures of buffers with FDLs have been considered. One is a feedforward architecture, consisting of parallel FDLs that have different lengths corresponding to desired delay times. A combination of input and output buffered switch [6] and multistage FDL buffer [7] were reported as feedforward architectures. The other is an architecture consisting of feedback-looped FDLs. It potentially provides infinite delay time if waveform distortion caused by loss, noise, dispersion etc., is managed to be compensated. However, the FDLs can provide only a restricted function of a finite delay time as buffers because the optical packet cannot be read out during the propagation in the FDLs.

In most of the proposed architectures, electrical processing for scheduling and management has been employed [8-12]. Although flexible control including quality of service (QoS) can be realized using such a control method, simple autonomous control is preferable for simple and low-power consumption buffering.

We have proposed an autonomous first-in-first-out (FIFO) buffer management system using all-optical sensing of packets [13]. Each of FDLs in the reported system stores a single packet. In this chapter, we describe architecture and operation of the buffering system. The buffering performances such as packet loss rate (PLR) and delay time are evaluated by numerical simulation.

2. Proposed Buffering System

2.1. Architecture of the Buffering System

The proposed buffering system consists of N parallel buffering modules and a combiner as shown in Fig.1. Each module can manage the buffering in autonomous fashion by exchanging the information signals that include utilization of output port as indicated by dashed lines. Packets forwarded by the buffering modules are transmitted to output port through an $N×1$ combiner.

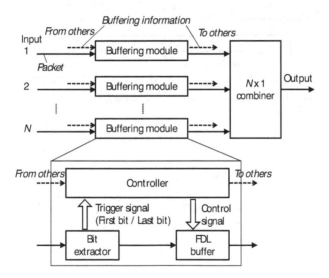

Figure 1. Schematic diagram of proposed buffering system.

The structure of a buffering module is also shown in inset of Fig. 1. It consists of a bit extractor, a controller, and an FDL buffer. The bit extractor creates trigger signal by detecting the first bit and the last bit of incoming packets. The first and last bits can be composed of specific coded bit patterns. Optical code-correlation processing can find the start and the end of the packet as the first and the last bits, respectively. The controller generates control signal autonomously by using the trigger signal. The FDL buffer stores and forwards packets by using the control signal.

2.2. Configuration of the Controller

Figure 2 shows the schematic diagram of the controller composed of four components. Controller A creates timing clock C_1 to be used to open the buffer for storing packets. Controller B creates 'store' signal which indicates the actually storing FDL in the buffer. Controller C creates another timing clock C_2 to be used to forward the already stored packets. Controller D creates 'forward' signal and the buffering information to other modules which indicates whether the buffer is now forwarding packets or not.

Figure 3 shows the configuration of controller A. Timing clock C_1 corresponds to the extracted first bit of incoming packets. The extracted last bit is not used in this case. However, it is reserved for future enhancement of the buffering system.

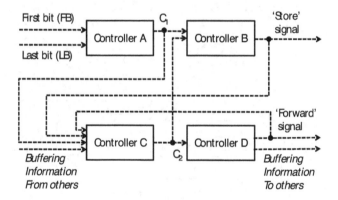

Figure 2. Schematic diagram of the controller.

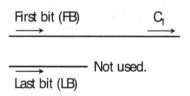

Figure 3. Configuration of controller A.

Figure 4 shows the configuration of controller B. It autonomously generates 'store' signals by processing C_1 and C_2. The number of 'store' signals is $M+1$ where M corresponds to the number of FDLs in the buffer. In order to indicate actual storing position, none or only one of the 'store' signals becomes on-state. When C_1 comes, position of on-state moves up from #1 to #($M+1$), namely, initially all off-state turns to #1-on, then moves to #2-on, #3-on, and so on. On the contrary, when C_2 comes, position of on-state moves down from #(M+1) to #1. There are some delay lines with delay time of T_{FIFO}-T_1 and T_1, where T_{FIFO} and T_1 are delay of single FDL in the buffer and 1-bit, respectively. Note that these components are expressed by some sort of logic circuits. Although it depends on the function and the performance such as operating speed, power consumption, and footprint, both electrical and optical logic circuits might be candidates to be employed.

Figure 5 shows the configuration of controller C. It autonomously generates C_2 by processing C_1, 'store' signal, buffering information signals from other modules, and 'forward' signal mentioned below. The operation of controller C is similar to the FDL buffer for packets. Therefore, we describe the detailed operation in the latter section about FDL buffer.

Figure 6 shows the configuration of controller D. It autonomously generates 'forward' signal by processing C_2. The 'forward' signal keeps on-state for a period of T_{FIFO} by using the flip-flop triggered by C_2.

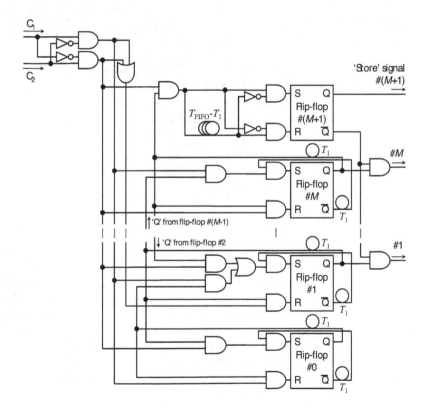

Figure 4. Configuration of controller B.

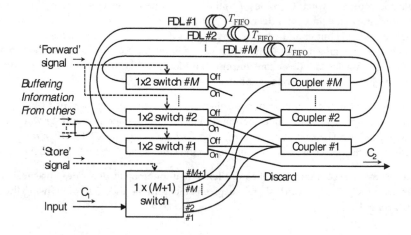

Figure 5. Configuration of controller C.

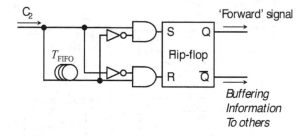

Figure 6. Configuration of Controller D.

2.3. Configuration of the FDL Buffer

The FIFO buffer consists of M parallel FDLs that have delay time of T_{FIFO}, a $1\times(M+1)$ input switch, $M1\times2$ output switches and couplers as shown in Fig. 7. The buffer stores and forwards packets by using 'store' and 'forward' signals, respectively. The stored position is determined by the state of 'store' signal. Namely, when k-th signal is on-state ($k=1,..,M$), incoming packets are switched to k-th FDL by the input switch. In case that $(M+1)$-th signal is on-state, then incoming packets will be discarded because all of the FDLs have already been occupied with other packets. The stored packets are forwarded to output by controlling the output switches. When the 'forward' signal is incident, all of the output switches move the stored packets to next neighbor FDLs. Note that T_{FIFO} is designed to be greater than or equal to the maximum packet length.

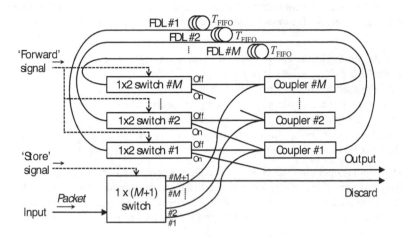

Figure 7. Configuration of the FDL buffer.

2.4. Operation Overview of Buffering

An example of timing chart for buffering process of a module is shown in Fig. 8.

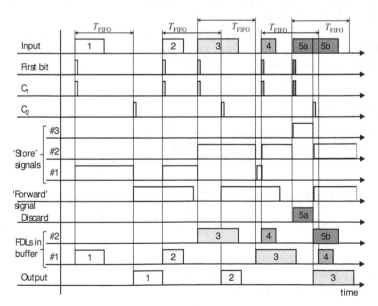

Figure 8. Timing chart example of a module of the buffer.

We assume that there are initially no packets stored in the FDLs, and then five packets are arriving sequentially with random timing and variable lengths. The number of FDLs M is assumed to be $M=2$.

When packet no.1 is incident, 'store' signal #1 turns on triggered by the first bit and C_1. The state of the 'store' signal is kept for a period of T_{FIFO}. When T_{FIFO} is expired, 'forward' signal turns on triggered by the C_2. Then, stored packets no.1 is forwarded to output. If another module has already been open for forwarding, the 'forward' signal of this module does not turn on in order to avoid collision between this module and the forwarding module. When packet no.3 is incident before the expiration period T_{FIFO} of packet no.2, it is stored into another FDL because each FDL is designed for storing only a single packet.

Although similar operation can be seen for following packets, packet no.5 is slightly different. The front part of it is discarded because all FDLs have already been occupied by other packets. When packet no.3 is forwarded, FDL #2 is open for storing. At the moment, the rear part of packet no.5 is stored into there. Note that the packet no.5 is therefore treated as a broken packet when it gets out from the buffer.

3. Computer Simulation

Two kinds of characteristics such as packet loss rate (PLR) and average delay time are investigated by computer simulation. We assume in the simulation that packets arrive randomly and have variable lengths from $L_{min}=10$ to $L_{max}=150$ bytes. We define load at input port by the ratio of the packet existence length to a unit length. For simplicity, operation speed of the composed devices, such as switching speed of some spatial switches and flip-flops, rise time of logic gates, are assumed to be much faster than bit-rate of arriving packets. Therefore, bit-rate is not specified in our simulation.

3.1. Packet Loss Rate

The PLR is verified with changing the number of FDLs M, length of each FDL L, number of input N, and the load. Because of a finite number of FDLs in the buffer system, overflow may occur when the load exceeds the capacity of the buffer, resulting in rejection of the overflowed packets. Even if the load is less than the capacity, collision of packets may occur when packets forwarded by some modules are simultaneously coming into the following combiner as shown in Fig.1. Therefore in the simulation, the overflow and the collision are both treated as loss of packet.

Figure 9 shows the PLR as a function of the load at module #1 with the number of FDLs M as a parameter. The number of modules is $N=2$. The load of the module #2 is set to 0.5. The length of each FDL is $L=L_{max}$. It is found that the PLR increases with load at module #1. Moreover, the PLR decreases when M increases.

Figure 10 shows the PLR as a function of the load at module #1 with the length of FDLs L as a parameter. The number of modules is $N=2$. The load of the module #2 is set to 0.5. The number of FDLs is $M=10$. It is found that the PLR increases both with load at module #1 and L. This is because incoming packets are separately stored in different FDLs, resulting in many FDLs are occupied with unused space remained. When L becomes long, the duration of occupation becomes long, and then it takes longer time to get out from the buffer. Therefore, it may cause the increase of packet loss because of the occupied FDLs.

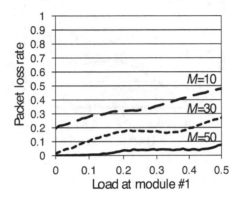

Figure 9. PLR with parameter M.

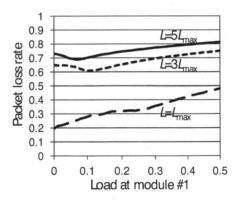

Figure 10. PLR with parameter L.

Figure 11 shows the PLR as a function of the load at module #1 with the load at module #2 as a parameter. The number of modules is $N=2$. The number of FDLs is $M=30$. The length of each FDL is $L=L_{max}$. It is found that the PLR increases with load at module #1 and #2.

Figure 12 shows the PLR as a function of the load at module #1 with the number of module N as a parameter. The number of FDLs is $M=30$. The length of each FDL is $L=L_{max}$.

Loads of modules other than #1 are all set to 0.3. It is found that the PLR increases with load at module #1 and N.

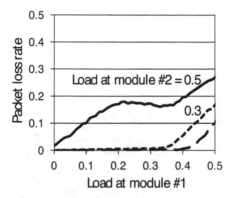

Figure 11. PLR with parameter load at module #2.

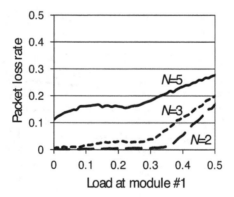

Figure 12. PLR with parameter N.

Figure 13 shows the breakdown of such numbers as input, output, discarded and broken packet in (a) module #1 and (b) module #2. Parameters are set to as follows; the number of modules is $N=2$, the number of FDLs is $M=30$, the length of each FDL is $L=L_{max}$, the load of module #1 is changed, and that of module #2 is 0.5 fixed. It is found from Fig. 13(a) that the number of output packets saturates up to 70 when the load at module #1 exceeds 0.3. This is because it starts overflow at that point, namely the number of discarded packets increases in proportion to the input. As a result, the PLR also increases. In case of module #2 shown in Fig. 13(b) that the fixed number of input initially exceeds its overflow limit, and so the number of discarded packets increases up to a certain value. Then, the number of output de-creases to the same level as that of module #1. Therefore, this 2×1 buffer puts the identical output priority to the two modules.

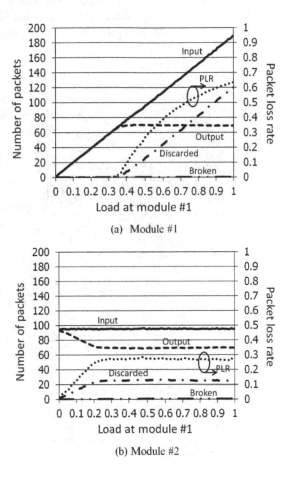

(a) Module #1

(b) Module #2

Figure 13. Breakdown of such numbers as input, output, discarded, and broken packet in each module.

3.2. Average Delay

Packets stored and forwarded through the buffer have been experienced a certain amount of delay determined mainly by the load and parameters M and L. We examine the average de-lay time by computer simulation.

Figure 14 shows the average delay as a function of the load at module #1 with the number of FDLs M as a parameter. The number of modules is $N=2$. The load of the module #2 is set to

0.5. The length of each FDL is $L=L_{max}$. It is found that the average delay increases with load at module #1 and M.

Figure 15 shows the average delay as a function of the load at module #1 with the length of each FDL L as a parameter. The number of modules is $N=2$. The load of the module #2 is set to 0.5. The number of FDLs is $M=10$. It is found that the average delay increases with load at module #1 and L.

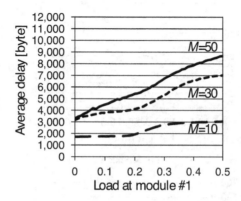

Figure 14. Average delay with parameter M.

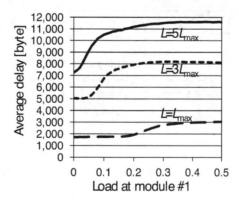

Figure 15. Average delay with parameter L.

Figure 16 shows distribution and moving average indicated by dots and solid curves, respectively, of delay time as a function of packet arrival time to each module with loads as parameters. The load of the module #1 is changed between 0.3, 0.5, and 0.7. The load of the module #2 is set to 0.5. The number of modules is $N=2$. The number of FDLs is $M=30$.

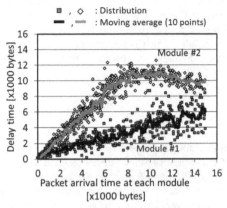

(a) Load at module #1=0.3, and load at module #2=0.5

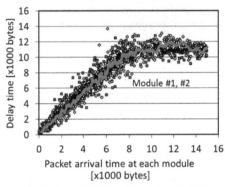

(b) Load at module #1=0.5, and load at module #2=0.5

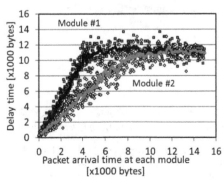

(c) Load at module #1=0.7, and load at module #2=0.5

Figure 16. Distribution of delay time at each module.

The length of each FDL is $L=L_{max}$. It is found that the delay time shows linear increase with packet arrival time because of the growth of buffer occupation. Moreover, the delay time shows saturation where the buffer occupation comes up to a maximum capacity. In addition, the heavily-loaded module, which corresponds to module #2 at Fig. 16(a) whereas module #1 at (c), shows faster increase and saturation than another module.

4. Conclusion

We have proposed an autonomous first-in-first-out buffer with capability of storing a single packet in each of FDLs. Characteristics of PLR and average delay have been investigated by numerical simulation. As a result, the PLR and the average delay have a trade-off relation at such parameters as number of FDL M and length of each FDL L. Therefore they should be determined by system demand. Smaller M and larger L can be options for implementing the system from viewpoints of footprint, power consumption, and avoid complicated control. Our future works include detailed investigation of buffering performance considering response time in switching and other constituent devices, and experimental verification.

Author details

Hiroki Kishikawa[1], Hirotaka Umegae[1], Yoshitomo Shiramizu[2], Jiro Oda[2], Nobuo Goto[1*] and Shin-ichiro Yanagiya[1]

*Address all correspondence to: goto@opt.tokushima-u.ac.jp

1 Department of Optical Science and Technology, The University of Tokushima, Japan

2 Department of Information and Computer Sciences, Toyohashi University of Technology, Japan

References

[1] Burmeister, E. F., Blumenthal, D. J., & Bowers, J. E. (2008, Mar). A comparisonof optical buffering technology. *Optical Switching and Networking*, 5(1), 10-18.

[2] Hunter, D. K., Chia, M. C., & Andonovic, I. (1998, Dec). Buffering in optical packet switches. *J. Lightwave Technol.*, 16(12), 2081-2094.

[3] Tucker, R. S., Ku, P., , C., & Chang-Hasnain, C. J. (2005, Dec). Slow-light optical buffers: Capabilities and fundamental limitations. *J. LightwaveTechnol.*, 23(12), 4046-4066.

[4] Baba, T. (2008, Aug). Slow light in photonic crystals. *Nature Photonics* [8], 465-473.

[5]   Fontaine, N. K., Yang, J., Pan, Z., Chu, S., Chen, W., Little, B. E., & Yoo, S. J. B. (2008, Dec). Continuously tunable optical buffering at 40Gb/s for optical packet switching networks. *J. Lightwave Technol.* [23], 3776-3783.

[6]   Yang, H., & Yoo, S. J. B. (2005, June). Combined input and output all-optical variable buffered switch architecture for future optical routers. *IEEE Photonics Technol. Lett.*, 17(6), 1292-1294.

[7]   Ogashiwa, N., Harai, H., Wada, N., Kubota, F., & Shinoda, Y. (2005, Jan). Multistage fiber delay line buffer in photonic packet switch for asynchronously arriving variable-length packets. *IEICE Trans. Commun.*, E88-B(1), 258-265.

[8]   Harai, H., & Murata, M. (2006, Aug). Optical fiber-delay-line buffer management in optical-buffered photonic packet switch to support service differentiation. *IEEE J. on Sel. Areas in Commun.*, 24(8), 108-116.

[9]   Wang, Z., Chi, C., & Yu, S. (2006, Aug). Time-slot assignment using optical buffer with a large variable delay range based on AVC crosspoint switch. *J. Lightwave Technol.*, 24(8), 2994-3001.

[10] Liew, S. Y., Hu, G., & Chao, H. J. (2005, Apr). Scheduling algorithms for shared fiber-delay-line optical packet switches- part I: The single-stage case. *J. Lightwave Technol.*, 23(4), 1586-1600.

[11] Shinada, S., Furukawa, H., & Wada, N. (2011, Dec). Huge capacity optical packet switching and buffering. *Optics Express*, 19(26), B406-B414.

[12] Kurumida, J., & Yoo, S. J. Ben. (2012, Mar/Apr). Nonlinear optical signal processing in optical packet switching systems. *IEEE J. Selected Topics in Quantum. Electron.*, 18(2), 978-987.

[13] Shiramizu, Y., Oda, J., & Goto, N. (2008, Aug). All-optical autonomous first-in-first-out buffer managed with carrier sensing of output packet. *Optical Engineering*, 47(8), 085006-1-8.

Chapter 5

Opto-Electronic Packaging

Ulrich H. P. Fischer

Additional information is available at the end of the chapter

1. Introduction

Future optical communication systems will use the high bandwidth of optical fiber in the optical frequency domain. Fast transmitter and receiver modules are basic elements of these systems, which are able now to transmit terabits/s of information via the fiber. Experiments with opto-electronic integrated circuits (OEICs) in laboratory test beds and field tests require a special packaging that respects system requirements such as high environmental stability and low optical insertion loss. Several concepts for fiber-chip coupling schemes had been proposed in the past. One of these is laser micro welding shown by[1], [2], [3], [4], [5], [6], [7]. This scheme is referring to the standard for high volume industrial manufacture. The investment costs for this laser welding equipment are considerably high. There are numerous proven techniques for aligning OEICs effectively. For laboratory use and rapid prototyping a flexible design is needed which is able to adapt different OEICs with changing dimensions to an existing module type.

In this chapter you will get general information what does opto-electronic packaging mean. Here fiber-chip coupling with basic coupling concepts will be illustrated. The different types of active adjusting and passive techniques are explained. Optical connectors play a very important role to interconnect different transmission systems. In passage 7 an overview of existing fiber connectors is shown. Afterwards, different optical module types for active and passive opto-electronic devices are described in details. Finally, the long-term stability of the modules must be tested and all reliability requirements for international test procedures are specified.

In its simplest arrangement, the packaging of OEICs involves the alignment and attachment of the light guiding areas of the OEIC and the optical fiber. At the beginning of this section, the basics of optical coupling theory with an introduction to optical mode fields and their matching by lenses is presented. Afterwards, a description of active and passive waveguide to waveguide coupling techniques will follow. Finally, optical connectors and the outline of different kinds of start of the art optical modules will be depicted followed by a short over-view of long-term stability tests.

At this point I would like to define the opto-electronic packaging which was given by [8]:

„*Opto-electronic packaging means working on the connection of opto-electronic integrated circuits to optical and electrical transmission lines and bias supply combined in a environmental stable housing.*"

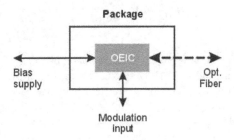

Figure 1. Basic package design for opto-electronic modules.

In the following several different technologies are listed which are essential to develop a new package:

1. RF-technique

2. Classical ray tracing optics & wave optics

3. Mechanical construction / CAD-design

4. Wire bond technique

5. Heat dissipating management / cooling

6. Communications engineering

7. Solid state physics

8. Micro systems design

9. Thick film circuits

10. Gluing, welding, soldering

2. Fiber-chip coupling

The behavior of the optical beam can normally be described by classical ray optical func-tions for lenses with focus length and focus point. If the dimensions of the optical beam come close to 1.5μm, which is the wavelength used in optical networks, the behavior of the beam must be described by wave optical functions.

$$p(r) = p(0) \times \exp \left\{ -2 \left[\frac{r}{w_0} \right]^2 \right\}$$

(1)

$2w_0$ =Mode field diameter(MFD)

$$w(z) = \sqrt{\left[w_0^2 + \left(\frac{z\lambda}{2n\pi w_0^2} \right)^2 \right]}$$

(2)

Here, the optical field within a wave-guide can be described nearly perfectly by a Gaussian intensity distribution, called p(r), which can be expressed with equation (1). If the wave travels within the waveguide, the mode field diameter is constant due to the combining function of the waveguide itself. At the end of the waveguide, the optical field is not guided and the field expands with increasing distance to the output facet. The expansion of the field can be calculated by equation (2). The point at which the intensity has fallen down to $1/e^2$ or 13.5% of the maximum intensity in radial direction, which is shown in figure 2, defines the mode field diameter.

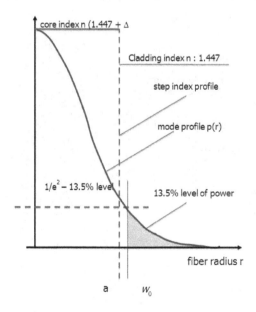

Figure 2. Intensity distribution of the optical mode field.

After leaving the waveguide, the optical mode field radius, which is half of the spot-size, expands with increasing distance to the facet. For distances of less than 200 µm the field distribution is called "near-field" and for larger distances "far-field". The angle where the intensity is fallen to $1/e^2$ or 13.5% of the maximum intensity is called "far-field angle" which corresponds to the near-field radius. These parameters are normally shown in the data sheets of laser diodes or LEDs.

For an efficient transfer of optical energy from the SMF and a laser diode wave guide, the mode profiles should "overlap" as much as possible which is described by [9]and is depicted in equation (3). The coupling efficiency ⊛ between two Gaussian beams can be expressed by means of the mode fields of laser diode w_{1d}, and fiber w_{SMF} and also as a function of lateral, angular and longitudinal misalignment between the two wave guides:

$$\eta = \kappa \cdot \exp\left[-\kappa\left\{\frac{x_0^2}{2}(1/w_{LD}^2 + 1/w_{SMF}^2) + \pi^2\theta^2\left[w_{LD}^2(z) + w_{SMF}^2\right]/2\lambda^2 - x_0\theta z/w_{AWG}^2\right\}\right] \tag{3}$$

Where

$$\kappa = 4w_{LD}^2 w_{SMF}^2 / \left[(w_{LD}^2 + w_{SMF}^2)^2 + \lambda^2 z^2 / \pi^2 n_{gap}^2\right] \tag{4}$$

With

λ- wavelength

N_{gap} – referactive index of medium between the waveguide and fiber

X_0 – lateral misalignment

θ- angular misalignment

Z – longitudinal misalignment

To measure the loss in decibel, the efficiency ⊛ must be multiplied 10 times by the logarithm$_{10}$ which is designated here as L:

$$L(\eta) = 10\log(\eta) \ \left[dB\right] \tag{5}$$

3. Basic coupling concepts

A comparison of the optical mode fields of the optical standard monomode fiber called SMF with a typical laser diode is shown in figure 3, where also the mode field diameter of a standard single mode fiber is depicted, respectively The properties of the SMF are standardized through the International Telecommunication Union [10].

The far field angle of the fiber is defined to a small value of 11.5°. A typical laser diode shows different values for lateral and vertical axis of 20° to 30° and 30° to 40°, respectively.

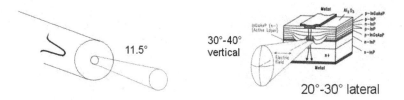

Figure 3. Far field of an optical fiber in comparison to the field of a laser diode.

If one compares the field parameters of fiber and laser diode a great mismatch can be found. Consequently, the optical coupling efficiency between these two devices is very low. The coupling loss between the two fields without additional mechanical misalignments can be calculated in decibel with Saruwatari's formula from equ. (3), which can be simplified into the formula (7):

$$Loss(R) \approx -10\log(R)\left[dB\right] \tag{6}$$

$$R = \frac{4}{\left\{\frac{w_1}{w_2} + \frac{w_2}{w_1}\right\}^2} \tag{7}$$

With this formula the mode field mismatches between the single mode components and the corresponding mismatch loss can be calculated to equ. (6), all lateral and angular misalignments of the fiber axis relative to the incident beam of the laser waveguide are set to zero.

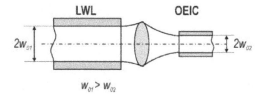

Figure 4. Mode field adaptation by an optical lens or lens system.

Different mode fields can be adapted by so called "mode field transformers". These devices are, for example, lenses or lens systems, which are shown in figure 4. With these devices, a nearly perfect coupling between the two wave-guides is possible. Due to limitations in costs, several lens configurations have been proposed for the optical coupling of laser diodes to single mode fibers. As shown in figure 5, simple ball lenses can be used to adapt the two mode fields. Coupling efficiencies of up to 30% are possible. A better approach is the use of graded index lenses or Selfoc-lenses. For a focusing lens, so-called half-pitch devices are used. Whereas quarter-pitch lenses are used to form parallel beams. These can reach efficiencies of 70%. Another approach is to form a lens at the end of the fiber, which is called fiber taper. With this device, efficiencies up to 90% have been achieved.

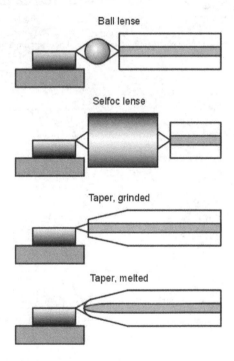

Figure 5. Mode field adaptation by several micro-optic solutions.

4. Adaptation of mode fields

As depicted in chapter 3, comparison of the optical fields of a butt ended standard single-mode fiber (SMF)and of edge emitting laser diodes shows a great mismatch. This mismatch is the reason for the very low coupling efficiency of approx. 15% for a butt ended fiber

This low efficiency can be overcome by a better adoption of the two optical mode fields with lenses. A coupling efficiency of more than 90% has been shown. Disadvantages occur at the handling of the parts because there are several parts including one ore two lenses, the fiber and the chip, which must be handled for optical alignment. The consequence is a rather costly of opto-electronic packaging.

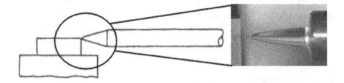

Figure 6. left side: Sketch of a melted fiber taper in front of an OEIC, right side: photograph.

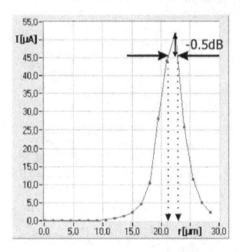

Figure 7. X and Y-axis: 2μm tolerance for 0.5 dB additional coupling loss.

As an example for integrated lens design, lenses can be made at the end of the fiber by melting the glass fiber and pulling it. This kind of fiber end is called fiber taper and works like a lens with typical diameters from 20 μm to 50 μm. In figure 6 you can see at the left side the chip facet and its wave guide and at the right side the melted fiber taper in front of the wave-guide. The fiber is additionally fixed into a metal cannula. With these tapered fibers a coupling efficiency of more than 50% can be realized. Unfortunately, a high precision fixation of better than 0.5 μm is necessary to mount the tapered fiber in front of the OEIC without additional losses. Therefore, the mechanical resolution of the coupling mechanism must be better than this value. The fixing procedure after coupling should not introduce additional displacements and must be stable enough to fix the coupling mechanism, which is important for a good long-term stability. The short working distance of 10 μm between fiber taper and laser which can be seen in the photograph is also dangerous for the life of the laser diode if it comes into contact with the with the fiber end. But there is only one low-priced device on the market, which makes this device very comfortable for use in small and very reasonably priced modules.

The tolerances for lateral and longitudinal fixing of the fiber taper in front of the opto-electronic circuit or OEIC are shown in figure 7 and figure 8. Both graphs show the distance in micrometers at the x-axis and a relative intensity of the coupling efficiency between the tapered fiber and the OEIC. You can see in figure 7 that within 2 micrometers the intensity will not be lower than 0.5 dB of the maximum intensity. For the longitudinal direction the tolerance is much greater: in this case 8μm, which can be seen in figure 8.

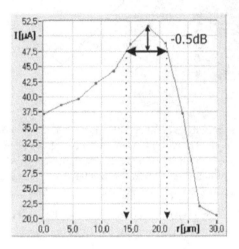

Figure 8. Z-axis with higher tolerance of 8μm.

Please remember: it is easier by a factor of four to perform the optical coupling in longitudinal direction in comparison to the lateral direction.

5. Active adjusting techniques

In optical packaging laboratories, fiber-chip coupling is performed within sub-micrometer precision in order to get a high coupling efficiency between the optical devices. Precision optical experiments depend upon reliable position stability. Vibration sources in and around the work are depicted in figure 9. Floors carry vertical vibrations in the range of 10 Hz to 30 Hz caused by people, traffic, seismic activity, and construction work. Tall buildings sway up to a meter in the wind, at frequencies from 1 Hz to 10 Hz. Machinery generates vibrations up to 200 Hz. Optical benches and their associated vibration isolating support systems provide a rigid and virtually vibration-free working surface that holds the components of an experiment in a fixed relative position. The legs support the tabletop: Air suspension mechanisms reduce practically all vibrations by two orders of magnitude.

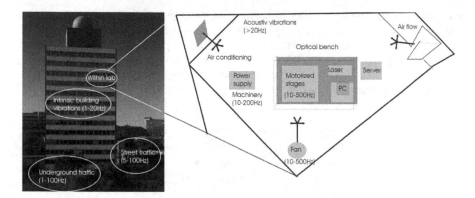

Figure 9. Influence of vibrations for micro positioning.

In order to align a fiber optically to another component, one has to move either the fiber, the component or both. Three linear and three angular motions are necessary to describe fully the motion and position of a solid body in space. The figure below identifies the six degrees of freedom using the common Cartesian frame of reference. When each of these degrees of freedom is singularly constrained by a hardware device, the device is labelled kinematic.

The device used to move a component in the linear x, y, or z direction is a translation stage. The device used to move a component in the angular θx, θy, or θz direction is a rotation stage.

Actuators are used to move the component on a translation stage to its desired position. There are three basic types of actuators used with precision stages: manual drives, stepper motor drives, and piezoelectric transducers. Manual drives and stepper-motor drives can move components over long distances, constrained only by the size of the manual drive or, in the case of stepper motors, the length of the lead screw. Piezoelectric transducers can move components over very short distances with nanometre precision. The range and resolution of the various drives and stage technologies is shown in figure 10 below.

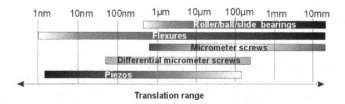

Figure 10. Translation ranges of available micro mechanical stages and screw systems.

For fiber-chip-alignment mostly all six degrees of freedom must be moved. Several commercial six axes motion systems have been developed with translatoric resolution of better then $0.02\mu m$ and angular resolution of better then one arc second. As an example, a mechanical/piezoelectric driven system with six degrees of freedom is shown in figure 11. Software tools are also included for automated coupling for one fiber and fiber arrays. Most of the software applications are available as a Labview virtual Instrument (VI).

Figure 11. Six axes Nano-positioning system.

6. Passive adjusting techniques

6.1. Flip-chip-technique

The flip-chip (FC) bond technology was developed in 1964 by IBM for high dense packages for hybrid modules. At that time it was called C4-technology or Controlled Collapse Chip Connection [11]. Its main goal was to replace the uneconomical wire bonds. The FC-technique allows the highest density of connections on the chip scale. For this reason this technique is a possible candidate for high volume production in the micro electronic industry.[1, 12] and [13] have shown that the FC-technique could meet the precision needed for optical fiber-chip coupling (± 3μm). Figure 12 shows the working principle of the optical FC-connection. The OEIC is connected bottom up to the substrate. Bonding is performed by a thermal reflow process which isdepicted in figure 13. The surface forces of the melted solder trek the OEIC into a preferred position, that differs very little from connection to connection. This is called "self-alignment". [14], [15]and [16] had developed a fluxless FC-bonding technique which allows a self-alignment of better than 1μm. Additionally, the short bond-distances promise very good RF-features up to 100GHz bandwidth. Today this progress allows the introduction of batch processing for the optical and electrical part of the optoelectronic packaging of OEICs. This benefit opens the market for high volume production of devices for optical communications systems that allows cost effective production of low budget products for the consumer market.

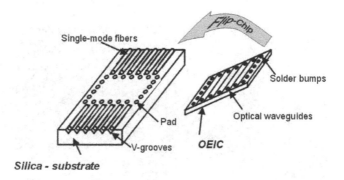

Figure 12. Flip-chip set-up.

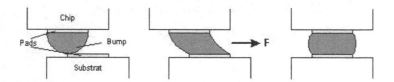

Figure 13. Flip-chip self-alignment.

6.2. Optical board technology

The progress in mechanical precision of the FC-bonds makes it possible to align one or multiple optical fibers direct to OEICs. Firstly, the OEIC will be FC-bonded as shown in figure 14. In the next step the fibers are inserted into V-grooves, fabricated by an anisotropic wet etching of the silica substrate. After insertion, the fiber must be fixed mostly by UV-hardened glue. Here more than 100 fibers can be arranged passively in one single fabrication step to the OEIC. An example of the connection of four lasers to an array of single mode fibers is shown in photograph 15.

In the next development step additional electrical amplifiers, multiplexers, modulators etc. can also be located on the substrate. This kind of hybrid integration is called optical motherboard or photonic lightwave circuit (PLC) depicted in figure 16. This type of integration is the most promising technique today for reaching an adequate price level of optical communications products for the consumer market.

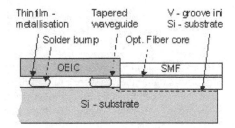

Figure 14. Fiber-chip connection with flip-chip.

Figure 15. Photograph of flip-chip bonded laser array(Heinrich-Hertz-Institute, Berlin).

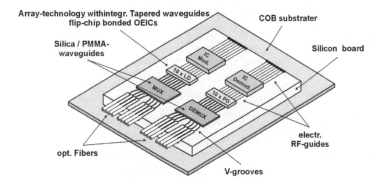

Figure 16. Concept for a complete optical motherboard [1], [17].

7. Optical connectors

7.1. Single fiber connectors

This chapter will give a summary of the optical connectors used today. There are several different types of connectors used for single mode and for multimode operation. Additionally, there are straight polished types and slant polished ones, which are used in high speed optical communication systems because of their high reflection loss characteristics. Further on, connectors are used in various applications including:

11. Polarization maintaining fibers

12. Matching gel/coating

13. Physical contact

14. Air gap

15. Automated light shutter function

16. Duplex connectors

Name	losses	Straight cut	Slant cut	Single (SM) Multi (MM) mode	Polarization maintaining	Durability (insertions)	Price
Mini BNC	0.21dB	-20dB		MM		"/500	low
ST	0.28dB	-20dB		SM,MM		"/500	med
FC/PC	0.35dB	-30dB		SM,MM		"/500	low
FC/APC	0.4dB		-55dB	SM		"/500	med
SMA	0.38dB	-20dB		MM		"/500	low
Radiall VFO	0.4dB	-30dB		SM	yes	"/250	high
	0.7dB		-55dB	SM		"/250	high
Radiall EC	0.2dB		-50dB	MM		"/250	med
	0.5dB		-60dB	SM	yes	"/250	med
Diamond E2000	0.18dB	-30dB		SM		"/1000	low
	0.18dB		-55dB	SM	yes	"/1000	low
SC	0.5dB	-30dB		SM		"/1000	low
	0.5dB		-60dB	SM	yes	"/1000	low
HRL-10	0.3dB		-60dB	SM		"/1000	High
LC-Duplex	0.2dB	-30dB		SM/MM		"/10.000	low

Table 1. Optical connectors summary.

All loss failure mechanisms that can be acknowledged at the fiber to fiber coupling are also detectable at connector-connector coupling. All possible losses are depicted in figure 17. Only highly precise mechanical feed and exact surface polishing can avoid high loss at the connection. Intrinsic losses can be avoided by using matching fibers, while extrinsic losses can be overcome by strong mechanical feed. Today feeder elements with better than 2 μm lateral deviation are commercially available. Polishing and cleaning the connector surface can avoid absorption and the scattering of the optical power. With the help of anti reflection coatings or angled surfaces, reflections can be (7°-8° degrees) overcome. All connectors are very similar in their mechanical structure. The fiber is fed through a ferrule made of ceramics, which centers the fiber. Than the ferrule is filled with UV-curing glue. After hardening, the end of the fiber is cut and polished. The outer diameter of the normally used ferrule is 2.5mm or 1.25mm. In figure 18 a cross section of a connector is depicted.

In the following, typical connectors used today are listed in table 1.

The most popular connector today is the FC/PC one direct followed by the Diamond E2000 and the very small SC connector. FC/PC-connectors (see figure 19) are mostly used in optical equipment but have the disadvantage to be easily soiled with dust and dirt. The E2000 is used by several Telecoms because of the integrated dust cover and beam shutter.

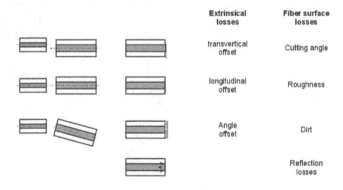

	Extrinsical losses	Fiber surface losses
	transvertical offset	Cutting angle
	longitudinal offset	Roughness
	Angle offset	Dirt
		Reflection losses

Figure 17. Loss mechanisms at connector end surface.

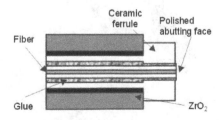

Figure 18. Cross section of an optical connector.

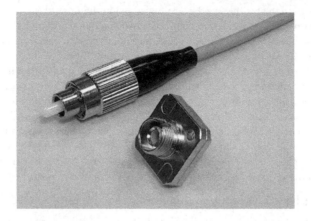

Figure 19. Optical FC/PC connector.

7.2. Multi-fiber connectors

These types of connectors are often used for the connections of mainframes or servers and at Fiber Distributed Data Interface (FDDI) links. Here the data were transmitted via several optical data links between the server stations with up to 10 Gbit/s per link. Also multi-sensor systems are using these kinds of connectors. Commercial available types are listed in table 2. The most commonly used connector is the MT one which is depicted in figure 20.

Figure 20. Multi-fiber MT-connector.

Name	Losses	Straight polished	Single (SM) Multi (MM) mode	durability (insertions)	User	Price
ESCON	0.5dB	-20dB	MM	"/500	LAN,Comp. Network	low
FDDI M	0.5dB 0,5dB	-20dB -20dB	MMSM	"/500"/500	WDM	med
MTConnect	0.3dB 0,5dB	-25dB -25dB	MMSM	"/200"/200	WDM	high

Table 2. Optical multi-fiber connectors.

8. Modules

This chapter will summarize several types of modules, which are used in commercial standard opto-electronic packages. These can be divided into four basic categories:

1. Transmitter (laser) -modules w/o cooling

2. Receiver (photodiode) -modules

3. Transceiver modules (Transmitter/Receiver)

4. Passive devices (sensor) -modules

Each category will be demonstrated by an example, but first some words about the required coupling method.

8.1. Fiber-chip coupling in modules

The performance of an optical coupling and the affordable operating expense are strongly dependent on the coupling device and the fiber used. The performance is directly correlated to the coupling efficiency. All commercial implementations are a trade off between cost and efficiency. To reach a very good efficiency you have to invest much manpower, which must be reflected in the price of the product. On the mass market only very low cost modules are available. That's the reason why here only modules with very low coupling efficiency can be found. The coupling of devices to multi mode fibers is less expensive than to single mode fibers, which are also priced lower than fiber-taper couplers. These are on the other hand are cheaper than lens-couplers, because of the time effort required to adjust additional coupling devices.

8.2. Multimode fiber coupling

The multimode fiber has an diameter of 50μm and a numerical aperture of 0.25 (15° aperture). Usually this type of coupling is used in photodiode modules where the fiber is directly connected to the photo-absorbing surface of the diode, which is depicted in figure 21.

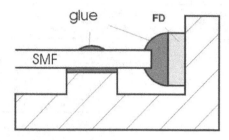

Figure 21. Direct attach of an optical fiber to a photo diode.

8.3. Singlemode-fiber coupling

In section 3 it has been demonstrated that a single mode butt fiber coupling to a Laser diode can have only 10 % efficiency. But with a lens system, which adapts the different optical mode fields, 50 % to 90 % coupling efficiency can be achieved (see figure 22). Table 4 gives a view of the mechanical aspects for coupling efficiencies with several coupling designs.

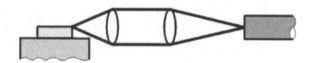

Figure 22. Fiber-chip coupling by a two-lens system.

Coupling technique (Laser-fiber)	Alignment precision (µm) for 1 dB add. loss at lateral/ longitudinaldisplacement	Coupling loss (dB)
Butt fiber (50µm)	15/50	7-10
Butt fiber (9µm)	2/20	7-10
System with one lens	0,5/5	3
Double lens system	0,5/5	1-3
Fiber taper	0,3/3	3-5

Table 3. Alignment precision for one dB additional coupling loss.

8.4. Transmitter and receiver modules

Transmitter modules for low cost applications are normally designed for simple butt fiber to chip coupling without temperature control of the emitting OEIC. Today more lensed coupling arrangements fixed by laser welding are often introduced. In figure 23, a coaxial coupled receiver module for high data rates of 40Gbit/s is depicted. This set-up is also used in low-priced transmitter modules for single mode operation.

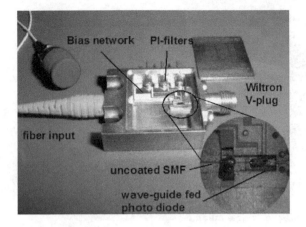

Figure 23. Receiver module.

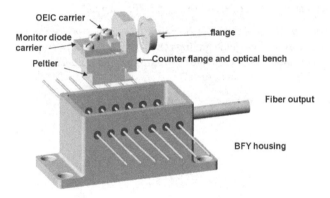

Figure 24. Temperature controlled laser module with fiber-taper coupling.

For high bit rate optical communications systems, cooled laser devices are needed. These modules are much more complicated in their mechanical set-up which is shown in figure 24. Here a tapered fiber was adjusted in front of the OEIC which is temperature stabilized by a Peltier cooler and a temperature sensor (thermistor) shown by [18].

8.5. Transceiver modules

These kind of modules are used in optical transmission systems where both terminals of the communications line can talk at the same time, which is called bi-directional communication. Transmitter and receiver functions must be integrated in these modules, which are shown in figure 25.

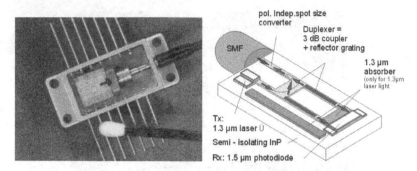

Figure 25. Monolithically integrated transceiver module.

8.6. Sensor and passive devices modules

These kind of modules are normally very easy to fabricate. Bragg sensors are used in a very wide spectrum of applications such as temperature sensors and strain gauge. The grating is centered into a metal or plastic tube and fixed with special glue.

Other passive devices use multiple fiber ports which can be combined in an array. Typical array devices are arrayed waveguide gratings (AWG)for multi wavelength optical transmission systems. These OEICs must be connected to up to 64 IO-ports at both chip sides as presented by [2], which can be seen in figure 27.

Typical housings are shown in figure 26.

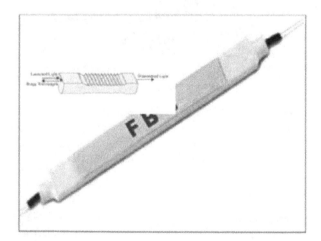

Figure 26. Fiber Bragg grating module.

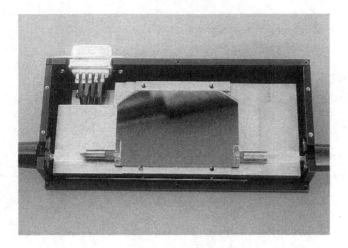

Figure 27. Arrayed waveguide grating module [19].

9. Reliability requirements

For application in optical networks, modules must be stable with respect to temperature changes and mechanical stresses. At present, there are several definite environmental and mechanical criteria for optical devices such as sensor and transmitter modules., which are investigated with reference to the [20] requirements. In the tests, insertion losses were measured online for each sample.

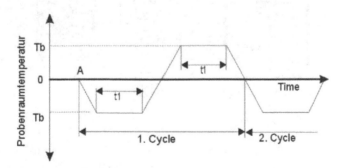

Figure 28. Temperature cycle test structure.

Temperature stress can be invoked into the modules by cycling the environmental temperature between a high temperature called TA and a low temperature TB that is depicted in figure 28. Additionally, relative air humidity can be increased up to 80 % or 95. The following institutions have developed the commonly used testing standards:

5.   DIN (Deutsche Industrie Norm, German Industrial Standard Organization) 40046

6.   MIL-STD (Military Standard/USA) 810/202

7.   IEC (International Engineering Committee) 60068-X

8.   Telcordia 6R-78, -326, -357, -468

Stress parameters	Tests
Climate	Cold, dry heat, dust and sand
	Low pressure
	Wheat heat at constant temperature
	Dry heat at cycling temperatures
	Solar radiation
Mechanical	Dropping, acceleration, vibrations
Chemical and biological	Corrosive atmosphere, growths of mold
Packaging and manufacturing	Welding, ultrasonic cleaning, mechanical strength of connector pins
div.	Sealing

Table 4. Environmental test parameters.

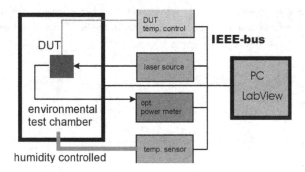

Figure 29. Environmental test set-up.

A typical set-up used for temperature testing is depicted in figure 29. Further, the device under test (DUT) is placed into a humidity controlled environmental test chamber. The temperature behavior of the opto-electronic module is mostly characterized by measuring the variation of optical output power. A plot of a typical temperature test between +15°C and +40°C is shown in figure 30. The temperature behavior of a laser module shows a maximum output variation of ±0.15 dB with temperature which is a real good result. Three cycles within 4 hours of measuring time were performed with temperature controlling of the OEIC shown in figure 30.

After thermal cycling, the test for mechanical shock and vibration stress was completed. Mechanical shock tests must be performed in all three Cartesian directions. The measured accelerations amounted to more than 200 g within 3 ms. The vibration tests where performed by a so called "shaker machine". The excitation of the module was measured with an acceleration sensor and a digital oscilloscope. The acceleration was controlled to be stronger than 16 g within a broad spectral bandwidth of 50-5000 Hz. Several tests figures can be run with the so called "shaker machine":

5. Sinusoidal acceleration1-1.000g

6. Noisy acceleration 1-10.000 Hz

7. Resonance test and -strain

8. Shock excitation 10 – 10.000g

To reach a certified test label for opto-electronic modules according to the Telcordia specifications, eleven modules must undergo the environmental and mechanical stress test. None of the tested specimens is allowed to show a failure. The strong requirements for the test procedures are only achieved by substantial preliminary testing of the modules.

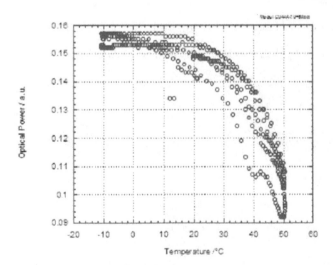

Figure 30. Temperature test of the laser module between -15°C and 50°C.

Sources of supply for environmental standard tests-

VDE-Verlag http://www.vde-verlag.de

DIN http://www.din.de

IEC-standards-shop http://www.iec-normen.de

Telcordiahttp://www.telecom-info.telcordia.com

10. Conclusion

We designed and fabricated a series of modules for one-sided and double-sided fiber-chip coupling for single mode and multimode fibers with simultaneous coupling of both chip sides by a new-patented set-up. Additional, we created passive and active modules with temperature control and multi fiber connections up to 16 fibers via fiber arrays. The modules have been tested in a reliability stress program between -40° C and +80°C and by a vibration shaker. Electrical modulation signals up to 50 GHz can be fed via RF connectors to the OEIC. The packages show good long-term stability and are well suited for rapid prototyping in laboratory environmentand high volume production.

Author details

Ulrich H. P. Fischer*

Address all correspondence to: ufischerhirchert@hs-harz.de

Harz University of Applied Sciences Friedrichstraße, Wernigerode

References

[1] Fischer, U. (2002). *Optoelectronic Packaging*, Berlin, VDE Verlag.

[2] Wild, T., & Stremitzer, S. (2007). *Digitale Wundanalyse mit W.H.A.T. (Wound Healing Analyzing Tool)*.

[3] http://www.mostnet.de, Available, http://www.mostnet.de/home/index.html.

[4] Homepage Aufbau- und Verbindungstechnik im HHI. Available, http://www.hhi.de/avt.

[5] Siegmund, S., Fischer-Hirchert, U. H. P., & Bauer, A. (2012). Technikgestützte Pflege-Assistenzsysteme und rehabilitativ-soziale Integration unter dem starken demografischen Wandel in Sachsen-Anhalt. Berlin. *in 5. Deutscher AAL-Kongress*.

[6] Reinboth, C., Fischer-Hirchert, U. H. P., & Witczak, U. (2012). Berlin. Technische Assistenzsysteme zur Unterstützung von Pflege und selbst-bestimmtem Leben im Alter- das ZIM-NEMO-Netzwerk TECLA, *in 5. Deutscher AAL-Kongress*.

[7]   Siegmund, S., Hirchert, A., Apfelbaum, A., & Fischer-Hirchert, U. H. P. (2012). Inno-vationslabor Technikakzeptanz. Görlitz. *in 13. Nachwuchs-Wissencshaftlerkonferenz mitteldeutscher Fachhochschulen*, 449-452.

[8]   Fischer, U. H. P. (2002). *Optoelectronic Packaging*.

[9]   Saruwatari, M., & Nawata, K. (1979). Semiconductor laser to single-mode fiber cou-pler. *Applied Optics*, 18, 1847-1856.

[10] ITU-G.652,. (1997). Characteristics of a single-mode optical fibre and cable. ed.

[11] http://www.fzk.de/imt/liga/d_index.html.

[12] http://www.izm.fhg.de/avt/kloeser.htm.

[13] Makiuchi, M., Norimatsu, M., Sakurai, T., Kondo, K., & Yano, M. (1993). Flip-Chip Planar Ga InAs/InP p-i-n Photodiode Array for Parralel Optical Transmission. *IEEE Photonics Technology Letters,*, 5, 518-520.

[14] Kuhmann, J. F. (1996). Untersuchungen zu einer flußmittelfreien und selbstjustieren-den Flip-Bondtechnologie für photonische Komponenten. TU, Berlin.

[15] Kuhmann, J. F., & Pedersen, E. H. (1998). Fluxless FC-soldering in O2 purged ambi-ent. Seattle, WA. *in 48TH IEEE Electronic Components & Technology Conference*, 256-258.

[16] Kuhmann, J. F., & Pech, D. (1996). In situ observation of the self-alignment during FC-bonding under vacuum with and without H2. *IEEE Photonics Technology Letters*, 8, 1665-1667.

[17] Fischer, U. H. P. (2006). *Optische Modenfeldadaption in photonischen Modulen der opti-schen Aufbau- und Verbindungstechnik*, Göttingen, Cuvillier-Verlag.

[18] http://www.imm-mainz.de, Available, http://www.imm-mainz.de/.

[19] Ehlers, H., Biletzke, M., Kuhlow, B., Przyrembel, G., & Fischer, U. H. P. (2000). Opto-electronic Packaging of Arrayed-Waveguide Grating Modules and Their Environ-mental Stability Tests. *Optical Fiber Technology*, 6, 344-356.

[20] Meyer, S., & Schulze, E. (2009). *Smart Home für ältere Menschen. Handbuch für die Prax-is*, Fraunhofer Irb verlag.

Chapter 6

A Method and Electronic Device to Detect the Optoelectronic Scanning Signal Energy Centre

Moisés Rivas, Wendy Flores, Javier Rivera,
Oleg Sergiyenko, Daniel Hernández-Balbuena and
Alejandro Sánchez-Bueno

Additional information is available at the end of the chapter

1. Introduction

In optoelectronic scanning, it has been found that in order to find the position of a light source, the signal obtained looks like a Gaussian signal shape. This is mainly observed when the light source searched by the optoelectronic scanning is punctual, due to the fact that when the punctual light source expands its radius a cone-like or an even more complex shape is formed depending on the properties of the medium through which the light is travelling. To reduce errors in position measurements, the best solution is taking the measurement in the energy centre of the signal generated by the scanner, see [1].

The Energy Centre of the signal concept considers the points listed below, see [2], in order to search which one of them represents the most precise measurement results:

- The Signal Energy Centre could be found in the peak of the signal.
- The Signal Energy Centre could be found in the centroid of the area under the curve Gaussian-like shape signal.
- The Signal Energy Centre could be found in the Power Spectrum Centroid.

The Energy Centre of the signal could be found by means of the optoelectronic scanner sensor output processing, through a computer programming algorithm, taking into account the points mentioned above, in a high level technical computing software for engineering and science like MATLAB. However, our contribution is a method and an electronic hardware to produce an output signal related to the Energy Centre in the optoelectronic scanning sensor, for applications in position measurements.

This method is based on the assumption that the signal generated by optical scanners for position measurements is a Gaussian-like shape signal. However, during experimentation it has been seen that the optoelectronic scanning sensor output is a Gaussian-like shape signal with some noise and deformation. This is due to some internal and external error sources like the motor eccentricity at low speed scanning, noise and deformation that could interfere with the wavelength of the light sources. Other phenomena could also affect such as of reflection, diffraction, absorption and refraction, producing a trouble that can be minimized by taking measurement in the energy centre of the signal.

The main interest of this chapter is to describe and explain a method to find the energy centre of the signal generated by optical scanners based on a dynamic triangulation, see [3], to reduce errors in position measurements.

2. Optoelectronic scanners for position measurements

Nowadays optoelectronic scanners are widely used for multiple applications; most of the position or geometry measuring scanners use the triangulation principle or a variant of this measurement method. There are two kinds of scanners for position measuring tasks: scanners with static sensors and scanners with rotating mirrors. Optical triangulation sensors with CCD or PSD are typically used to measure manufactured goods, such as tire treads, coins, printed circuit boards and ships, principally for monitoring the target distance of small, fragile parts or soft surfaces likely to be deformed if touched by a contact probe.

2.1. Scanners with position triangulation sensors using CCD or PSD

A triangulation scanner sensor can be formed by three subsystems: emitter, receiver, and electronic processor as shown in figure 1. A spot light is projected onto the work target; a portion of the light reflected by the target is collected through the lens by the detector which can be a CCD,CMOS or PSD array. The angle(α) is calculated, depending on the position of the beam on the detectors CCD or PSD array, hence the distance from the sensor to the target is computed by the electronic processor. As stated by Kennedy William P. in [4], the size of the spot is determined by the optical design, and influences the overall system design by setting a target feature size detection limit. For instance, if the spot diameter is 30 µm, it will be difficult to resolve a lateral feature <30 µm.

Many devices are commonly utilized in different types of optical triangulation position scanners and have been built or considered in the past for measuring the position of light spot more efficiently. One method of position detection uses a video camera to electronically capture an image of an object. Image processing techniques are then used to determine the location of the object. For situations requiring the location of a light source on a plane, a position sensitive detector (PSD) offers the potential for better resolution at a lower system cost[5]. However, there are other kinds of scanners used commonly in large distances measurement or in structural health monitoring tasks, these scanners will be explained in the next section.

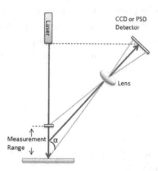

Figure 1. Principle of Triangulation.

2.2. Scanners with rotating mirrors and remote sensing

In the previous section, we described the operational principle of scanners for monitoring the distance of small objects, now we will describe the operational principle of scanners with rotating mirrors for large distances measurement or in structural health monitoring tasks.

There are two main classification of optical scanning: remote sensing and input/output scanning. Remote sensing detects objects from a distance, as by a space-borne observation platform. For example an infrared imaging of terrain. Sensing is usually passive and the radiation incoherent and often multispectral. Input / output scanning, on the other hand, is local. A familiar example is the document reading (input) or writing (output).The intensive use of the laser makes the scanning active and the radiation coherent. The scanned point is focused via finite-conjugate optics from a local fixed source, see [6].

In remote sensing there is a variety of scanning methods for capturing the data needed for image formation. These methods may be classified into framing, push broom, and mechanical. In the first one, there is no need for physical scan motion since it uses electronic scanning and implies that the sensor has a two-dimensional array of detectors. At present the most used sensor is the CCD and such array requires an optical system with 2-D wide-angle capability. In push broom methods a linear array of detectors are moved along the area to be imaged, e. g. airborne and satellite scanners. A mechanical method includes one and two dimensional scanning techniques incorporating one or multiple detectors and the image formation by one dimensional mechanical scanning requires the platform with the sensor or the object to be moved in order to create the second dimension of the image.

In these days there is a technique that is being used in many research fields named Hyperspectral imaging (also known as imaging spectroscopy). It is used in remotely sensed satellite imaging and aerial reconnaissance like the NASA's premier instruments for Earth exploration, the Jet Propulsion Laboratory's Airborne Visible-Infrared Imaging Spectrometer (AVIRIS) system. With this technique the instruments are capable of collecting high-dimensional image data, using hundreds of contiguous spectral channels, over the same area on the surface of the Earth, as shown in figure 2. where the image measures the reflected radiation in the wavelength region from 0.4 to 2.5 μm using 224 spectral channels, at nominal spectral resolution of 10 nm. The wealth of spectral information provided by the latest generation hyperspectral sensors has opened ground breaking perspectives in many applications, including environmental modelling and assessment; target detection for military and defence/security deployment; urban planning and management studies, risk/hazard prevention and response including wild-land fire tracking; biological threat detection, monitoring of oil spills and other types of chemical contamination [7].

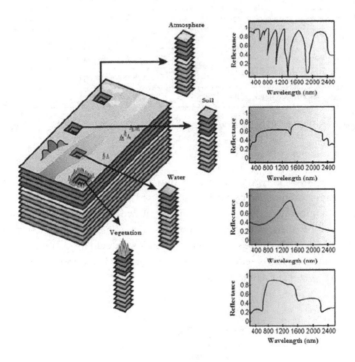

Figure 2. The concept of hyperspectral imaging illustrated using NASA's AVIRIS sensor [7].

While remote sensing requires capturing passive radiation for image formation, active input/output scanning needs to illuminate an object or medium with a "flying spot, " derived typically from a laser source. In Table 1, we listed some examples divided into two principal functions: input (when the scattered radiation from the scanning spot is detected) and output (when the radiation is used for recording or displaying). Therefore, we can say that in input scanning the radiation is modulated by the target to form a signal and in the output scanning it is modulated by a signal.

Input / Output Scanning	
Input	**Output**
Image scanning / digitising	Image recording / printing
Bar-code reading	Colour image reproduction
Optical inspection	Medical image outputs
Optical character recognition	Data marking and engraving
Optical data readout	Micro image recording
Graphic arts camera	Reconnaissance recording
Scanning confocal microscopy	Optical data storage
Colour separation	Phototypesetting
Robot vision	Graphic arts platemaking
Laser radar	Earth resources imaging
Mensuration (Measurement)	Data / Image display

Table 1. Examples of Input / Output Scanning.

2.2.1. Polygonal scanners

These scanners have a polygonal mirror rotating at constant speed by way of an electric motor and the radiation received by the lens is reflected on a detector. The primary advantages of polygonal scanners are speed, the availability of wide scan angles, and velocity stability. They are usually rotated continuously in one direction at a fixed speed to provide repetitive unidirectional scans which are superimposed in the scan field, or plane, as the case may be. When the number of facets reduces to one, it is identified as a monogon scanner, figure 3 illustrates an hexagonal rotating mirror scanner.

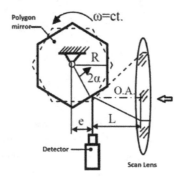

Figure 3. Polygon scanner.

2.2.2. Pyramidal and prismatic facets

In these types of scanners, the incoming radiation is focused on a regular pyramidal polygon with a number of plane mirrors facets at an angle, rather than parallel, to the rotational axis. This configuration permits smaller scan angles with fewer facets than those with polygonal mirrors. Principal arrangements of facets are termed prismatic or pyramidal. The pyramidal arrangement allows the lens to be oriented close to the polygon, while the prismatic configuration requires space for a clear passage of the input beam.

2.2.3. Holographic scanners

Almost all holographic scanners comprise a substrate which is rotated about an axis, and utilize many of the characterising concepts of polygons. An array of holographic elements disposed about the substrate serves as facets, to transfer a fixed incident beam to one which scans. As with polygons, the number of facets is determined by the optical scan angle and duty cycle, and the elemental resolution is determined by the incident beam width and the scan angle. In radially symmetric systems, scan functions can be identical to those of the pyramidal polygon. Meanwhile there are many similarities to polygons, there are significant advantages and limitations.

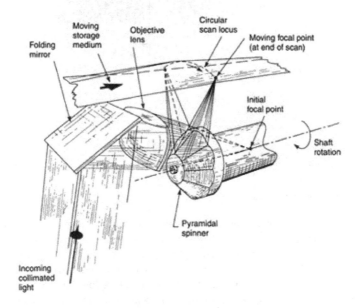

Figure 4. Polygonal scanner (From http://beta.globalspec.com/reference/34369/160210/chapter-4-3-5-4-scanner-devices-and-techniques-postobjective-configurations).

2.2.4. Galvanometer and resonant scanners

To avoid the scan non uniformities which can arise from facet variations of polygons or holographic deflectors, one might avoid multifacets. Reducing the number to one, the polygon becomes a monogon. This adapts well to the internal drum scanner, which achieves a high duty cycle, executing a very large angular scan within a cylindrical image surface. Flat-field scanning, however, as projected through a flat-field lens, allows limited optical scan angle, resulting in a limited duty cycle from a rotating monogon. If the mirror is vibrated rather than rotated completely, the wasted scan interval may be reduced. Such components must, however, satisfy system speed, resolution, and linearity. Vibrational scanners include the familiar galvanometer and resonant devices and the least commonly encountered piezo-electrically driven mirror transducer as shown in figure 5.

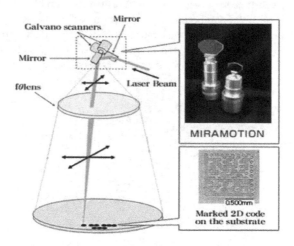

Figure 5. Galvanometer scanner (From http://www.yedata.com).

2.2.5. 45° cylindrical mirror scanner

Optical scanning systems can use coherent light emitting sources, such as laser or incoherent light sources like the lights of a vehicle. In the use of laser as light emitting source, the measurements are independent of environment lighting, so it is possible to explore during day and night, however, there are some disadvantages such as the initial cost, the hazard due to its high energy output, and that they cannot penetrate dense fog, rain, and warm air currents that rise to the structures, interfering the laser beam, besides, it is difficult to properly align the emitter and receiver. A passive optical scanning system for SHM can use conventional light emitting sources placed in a structure to determine if its position changes due to deteriorating. Figure 6 illustrates a general schematic diagram with the main elements of the optical scanning aperture used to generate the signals to test the proposed method.

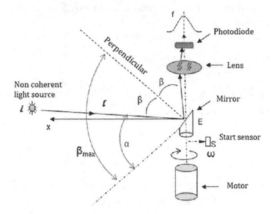

Figure 6. Cylindrical Mirror Scanner.

The optical system is integrated by the light emitter source set at a distance from the receiver; the receiver is compound by the mirror E, which spins with an angular velocity ω. The beam emitted arrives with an incident angle β with respect to the perpendicular mirror, and is reflected with the same angle β, according to the reflecting principle (C L. Wyatt, 1991) to pass through a lens that concentrates the beam to be captured by the photodiode, which generates a signal "f" with a shape similar to the Gaussian function. When the mirror starts to spin, the sensor "s" is synchronized with the origin generating a pulse that indicates the starting of measurement that finishes when the photodiode releases the stop signal. This signal is released when the Gaussian signal energetic centre has been detected.

Figure 7 shows that light intensity increments in the centre of the signal generated by the scanner. The sensor "s" generates a starting signal when $t\alpha = 0$, then the stop signal is activated when the Gaussian function geometric centre has been detected.

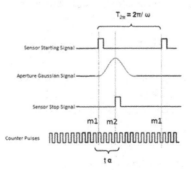

Figure 7. Signal generated by a 45° cylindrical mirror scanner.

The distance T2π is equal to the time between m1 and m1, that are expressed by the code N2π as defined in equation 1.

$$N_{2\pi} = T2\pi \cdot f_0 \tag{1}$$

On the other hand, the time tα is equal to the distance between m1 andm2, could be expressed by the code defined in equation 2.

$$N\alpha = t_{\alpha} \cdot f_0 \tag{2}$$

Where f0 is a standard frequency reference. With this consideration the time variable could be eliminated from equation 2obtaining equation 3,see [8].

$$\alpha = 2\pi \cdot N_{\alpha} / N_{2\pi} \tag{3}$$

3. Scanner sensors

All detectors (sensors) act as transducer that receive photons and produce an electrical response that can be amplified and converted into a form of relevant parameters to handle the input data for results interpretation. Among relevant parameters we can find spectral response, spectral bandwidth, linearity, dynamic range, quantum efficiency, noise, imaging properties and time response. Photon detectors respond directly to individual photons. Absorbed photons release one or more bound charge carriers in the detector that modulates the electric current in the material and moves it directly to an output amplifier. Photon detectors can be used ina spectral band width from X-ray and ultraviolet to visible and infrared spectral regions. We can classify them as analogue waveform output and image detectors, however, another type of classification is also possible but we will only describe this type of sensors in this section.

Analogue waveform output detectors

Analogue waveform output detectors are used as an optical receiver to convert light into electricity. This principle applies to photo detectors, phototransistors and other detectors as photovoltaic cells, and photo resistance, but the most widely used today in position measuring process are the photodiode andthe phototransistors.

3.1.1. Photodiode

The photodiode could convert light in either current or voltage, depending upon the mode of operation. A photodiode is based on a junction of oppositely doped regions (pn junction) in a sample of a semiconductor. This creates a region depleted of charge carriers that results in high impedance. The high impedance allows the construction of detectors using silicon and germanium to operate with high sensitivity at room temperatures.

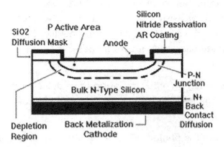

Figure 8. Cross section of a typical silicon photodiode.

A cross section of a typical silicon photodiode is shown in the figure. 8. N type silicon is the starting material. A thin "p" layer is formed on the front surface of the device by thermal diffusion or ion implantation of the appropriate doping material (usually boron). The interface between the "p" layer and the "n" silicon is known as a pn junction. Small metal contacts are applied to the front surface of the device and the entire back is coated with a contact metal. The back contact is the cathode; the front contact is the anode. The active area is coated with silicon nitride, silicon monoxide or silicon dioxide for protection and to serve as an anti-reflection coating. The thickness of this coating is optimized for particular irradiation wavelengths.

In semiconductors whose bandgaps permit intrinsic operation in the 1-15μm, a junction is often necessary to achieve good performance at any temperature. Because these detectors operate through intrinsic rather than extrinsic absorption, they can achieve high quantum efficiency in small volumes. However, high performance photodiodes are not available at wavelengths longer than about 1.5μm because of the lack of high-quality intrinsic semiconductors with extremely small bandgaps. Standard techniques of semiconductor device fabrication allow photodiodes to be constructed in arrays with many thousands, even millions, of pixels. Photodiodes are usually the detectors of choice for 1-6μm and are often useful not only at longer infrared wavelengths but also in the visible and near ultraviolet.

The photodiode operates by using an illumination window, which allows the use of light as an external input. Since light is used as an input, the diode is operated under reverse bias conditions. Under the reverse bias condition the current through the junction is zero when no light is present, this allows the diode to be used as a switch or relay when sufficient light is present.

Photodiodes are mainly made from gallium arsenide instead of silicon because silicon creates crystal lattice vibrations called phonons when photons are absorbed in order to create electron-hole pairs. Gallium arsenide can produce electron-hole pairs without the slowly moving phonons; this allows faster switching between on and off states and Ga As also is more sensitive to the light intensity. Once charge carriers are produced in the diode material, the carriers reach the junction by diffusion.

Photodiodes are similar to regular semiconductor diodes except that they may be either exposed to detect vacuum UV or X-ray or packaged with a windows or optical fibre connection to allow light to reach the sensitive part of the device. Many diodes designed to use specifically as a photodiode use a PIN junction rather than a p-n junction, to increase the speed of response [9].

Spectral response: The wavelength of the radiation to be detected is an important parameter. As shown in figure 9, silicon becomes transparent to radiation of a wavelength longer than 1100 nm.

Linearity: Current output of the photodiode is very linear with radiant power throughout a wide range. Nonlinearity remains below approximately 0.02% up to 100mA photodiode current. The photodiode can produce output currents of 1mA or greater with high radiant power, but nonlinearity increases to a certain percent in this region. This excellent linearity at high radiant power assumes that the full photodiode area is uniformly illuminated. If the light source is focused on a small area of the photodiode, nonlinearity will occur at lower radiant power.

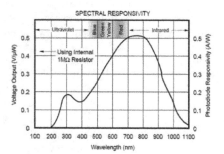 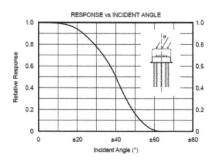

Figure 9. Spectral responsivity and response vs. incident angle of a photodiode.

Dynamic Range: Dynamic response varies with feedback resistor, using 1M resistor, the dynamic response of the photodiode can be modelled as a simple R/C circuit with a − 3dB cut off frequency of 4kHz. This yields a rise time of approximately 90μs (10% to 90%). See figure 10.

Noise: The noise performance of a photo detector is sometimes characterized by Noise Effective Power (NEP). This is the radiant power which would produce an output signal equal to the noise level. NEP has the units of radiant power (watts). The typical performance curve "Noise Effective Power vs. Measurement Bandwidth" shows how NEP varies with RF and measurement bandwidth.

Imagining Properties: The output is measured in voltage thru time, imaging like a Gaussian-like signal shape [10].

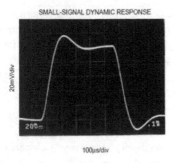 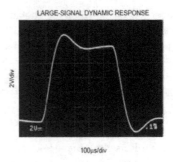

Figure 10. Small and large signal dynamic response of a photodiode.

3.1.2. Phototransistors

The Phototransistor is similar to the photodiode except for an n-type region added to the photodiode configuration. The phototransistor includes a photodiode with an internal gain. A phototransistor can be represented as a bipolar transistor that is enclosed in a transparent case so that photons can reach the base-collector junction. The electrons that are generated by photons in the base-collector junction are injected into the base, and the photodiode current is then amplified by the transistor's current gain β (or hfe). Unlike photodiode phototransistor cannot detect light any better, it means that they are unable to detect low levels of light. The drawback of a phototransistor is the slower response time in comparison to a photo diode. If the emitter is left unconnected, the phototransistor becomes a photodiode [11].

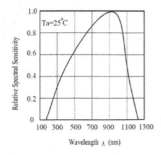 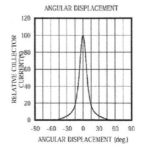

Figure 11. Relative spectral sensitivity and collector current vs. angular displacement of a phototransistor.

3.2. Image sensors

Nowadays image sensors are recognized as the most advanced technology to record electronic images.

These sensors are based on the photoelectric effect in silicon. When a photon of an appropriate wavelength (in general between 200 and 1000 nm) hits silicon, it generates an electron-hole pair. If an electric field is present, the electron and the hole are separated and charge can accumulate, proportional to the number of incident photons, and therefore the scene imaged onto the detector will be reproduced if a proper X-Y structure is present. Each basic element, defining the granularity of the sensor, is called a pixel (picture element) [12].

3.2.1. CCD sensor (charge coupled device)

This sensor is used in scanners to capture digital Images. Typically, it is an array to perform the scanning row by row, scanning one horizontal row pixel at a time, moving the scan line down with a carriage motor. The scanners that use this CCD sensor cell use an optical lens, often like a fine camera lens, and a system of mirrors to focus the image onto the CCD sensor cells.

The CCD sensor cell is an analogue device, when light strikes the chip, it is held as a small electrical charge in each photo sensor. The charges are converted to voltage, one pixel at a time, as they are read from the chip. An additional circuitry is also required to convert analogue to digital signal to produce an image as shown in figure 12.

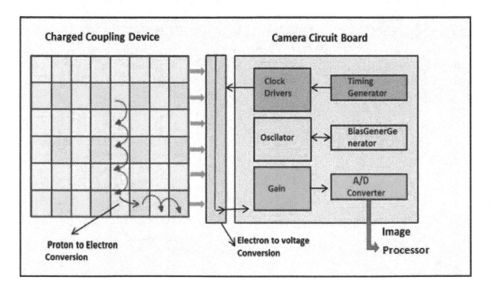

Figure 12. CCD operating principle.

The basic concept o CCDs is a simple series connection of Metal-Oxide-Semiconductor ca-pacitors (MOS capacitors). The individual capacitors are physically located very close to each other. The CCD is a type of charge storage and transport device: charge carriers are stored on the MOS capacitors and transported. To operate the CCDs, digital pulses are ap-plied to the top plates of the MOS structures. The charge packets can be transported from one capacitor to its neighbour capacitor. If the chain of MOS capacitors is closed with an out-put node and an appropriate output amplifier, the charges forming part of a moving charge packet can be translated into a voltage and measured at the outside of the device. The way the charges are loaded into the CCDs is application dependent.

The advantages of CCDs are size, weight, cost, power consumption, stability and image quality (low noise, good dynamic range, and colour uniformity). A disadvantage is that it is susceptible to vertical smear from bright light sources when the sensor is overloaded [13].

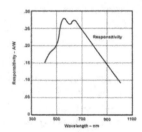 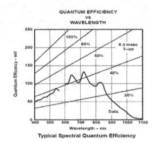

Figure 13. Responsivity and quantum efficiency vs. wavelength of a CCD.

As light enters the active photo sites in the image area, electron hole pairs are generated and the electrons are collected in the potential wells of the pixels. The wells have a finite charge storage capacity determined by the pixel design [14].

3.2.2. CMOS sensor (complementary metal oxide semiconductor)

This is an active pixel sensor or image sensor fabricated with an integrated circuit that has an array of pixel sensors, each pixel containing both a CMOS component and an ac-tive amplifier. Extra circuitry next to each photo sensor converts the light energy to a voltage and additional circuitry is also required to convert analogue to digital signal. In CMOS sensor the incoming photons go through colour filters, then through glass layers supporting the metal interconnect layers, and then into silicon, where they are absorbed, exciting electrons that then travel to photodiode structures to be stored as signal. These are commonly used in cell phone cameras and web cameras. They can potentially be implemented with fewer components, use less power, and/or provide faster readout than CCDs, scaling to high resolution formats. CMOS sensors are cheaper to manufac-ture than CCD sensors. However, a disadvantage is that they are susceptible to unde-sired effects that come as a result of rolling shutter [15].

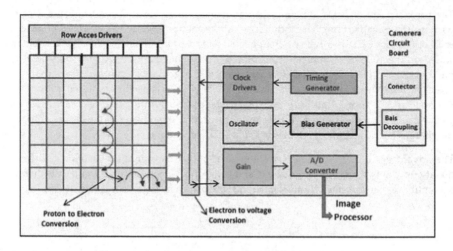

Figure 14. CMOS sensor.

Typically the sensitivity of the sensor is evaluated based on the quantum efficiency, QE, or the chance that one photon generates one electron in the sensor at a given wavelength. This is a good indicator. This gives the minimum amount of light you can see.In general, CMOS sensors have a higher QE in the sensitivity due to their design structure, and this can be further optimized by producing the sensor using a thicker epitaxial layer (shown as CMOS 1-b in Figure 15 below).Hence, at 800nm, the CMOS sensor with the thicker epitaxial layer has the best QE.

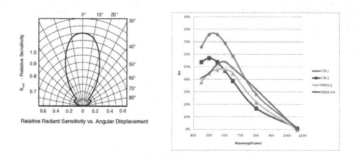

Figure 15. Relative radiant sensitivity vs. angular displacement and CCD vs. CMOS sensitivity.

3.2.3. Position sensing detector

Other device widely used as triangulation position sensor is the PSD (Position Sensing Detector), which converts an incident light spot into continuous position data(figure 16)and is more accurate and faster than CCD because the PSD is a continuous sensor, while CCD is a matrix of dots switched on and off and its resolution depends on how many dots are located on the sensor. Typically a linear CCD has 1024 or 2048 dots.

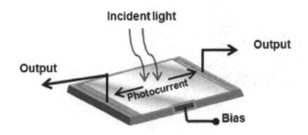

Figure 16. PSD Operating principle.

PSD has an infinite resolution because it is a continuous sensor, therefore the digital resolution of a PSD depends not on the PSD itself. Alignment sensors using CCDs have to be programmed to do multiple measurements at every step to improve accuracy and to lower noise because linear CCDs have a low resolution. To have the same accuracy of a PSD, CCD should perform no less than 32 measurements and hence calculate the average measurement. However, CCD is generally preferred to PSD because PSD needs an expensive circuit design including Analogue-to-Digital conversion [17].

4. Typical optoelectronic scanners signals

Different shapes of signals are generated during an optical scanning process, depending on the kind of light source and the sensor of the scanner. Some precision semiconductor optical sensors like CCD or PSD produce output currents related to the "centre of mass" of light incident on the surface of the device See [18]. All light registered by the CCD or PSD originates an ideal signal shape as shown in figure 17(Image credit: Measurecentral.com).

A typical position measuring process includes an emitter source of light, as a laser diode or an incoherent light lamp and the position sensitive detector like CCD or PSD as a receiving device, which collects a portion of the back-reflected light from the target. The position of the spot on the PSD is related to the target position and the distance from the source, see[19].However, the real photon distribution on the sensor depends on the characteristic dimensions related to the diffraction pattern of the light in the space. Common examples of signals generated by the light registered on a CCD camera appear in a study about super-resolution by spectrally selective imaging, shown in figure 18 (Image credit: A.M. van Oijen and J.Köhler).

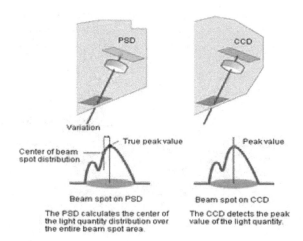

Figure 17. Ideal photon distribution on CCD and PSD sensors.

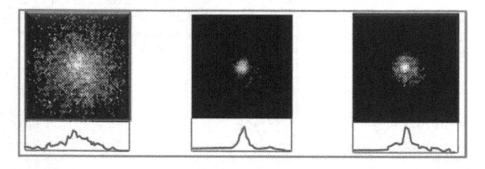

Figure 18. Real photon distributions as a function of the detector for different diffraction patterns.

Based on A.M. van Oijen and J.Köhler study, we can observe that the spatial distribution function of light has an Airy-function-like shape, see [20]. It is well known that CCD, CMOS and SPD use the light quantity distribution of the entire beam spot entering the light receiving element to determine the beam spot centre or centroid and identifies this as the target position. However, they are not the only sensors that generate a similar Gaussian-like shape, there are still a lot of sensors to be further investigated. For example, a simple photodiode can also originate a similar Gaussian-like shape, when it is used as a sensor on a scanner with a rotating mirror, [21]. Figure 19 below illustrates a hypothetical spot model, and attempts to explain how the signal is created by the photodiode on a scanner with a rotating mirror.

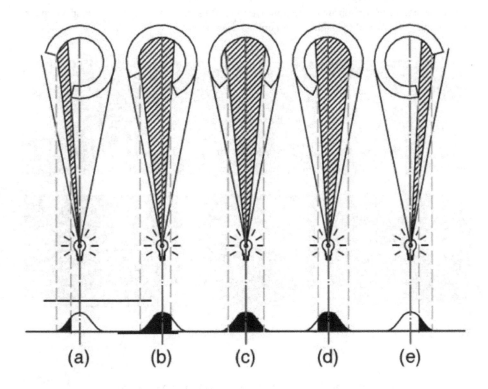

Figure 19. Principle of electrical signal formation during rotational scanning.

In this case, the signal created, as a similar Gaussian-like shape, goes up(Fig. 19, a) and falls down (Fig. 19, e), and a fluctuating activity takes place around its maximum area in figs. 19 (b-d). As we mentioned before, in a real practice the signal becomes noisy, see [22]. The experiment recently developed by Rivas M. and Flores W., with the scanner shown in Figure 6, for angular position measuring and using an incoherent light source and a simple photodiode, validated the model shown in Figure 20. During experimentation, it has been observed that the optoelectronic scanning sensor (photodiode) output is a Gaussian-like shape signal with some noise and deformation. This is due to some internal and external error sources like the motor eccentricity at low speed scanning, noise and deformation that could interfere with the wavelength of the light sources. Other phenomena could also affect, though, such as reflection, diffraction, absorption and refraction, producing a as seen in Figure 20.

As we can see, the photodiode signal originates a similar function to a CCD, consequently, it is possible to enhance the accuracy measurements in optical scanners with a rotating mirror, using a method for improving centroid accuracy by taking measurement in the energy centre of the signal. In the following section, we propose a new method and its respective circuit to find the centre of the signal by an optical scanner.

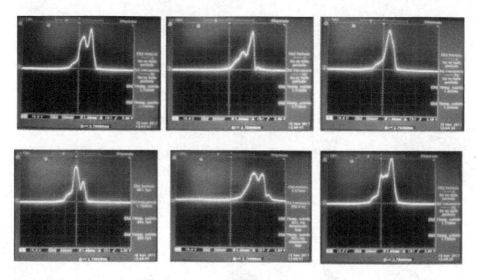

Figure 20. Scanning sensor Gaussian like shape, measurements at different angular positions of the light source.

5. Signal processing methods to locate signal energy centre

In the previous sections, we described different sensors and scanners that produce output currents related to the "the centre of mass" of the light incident in the surface of the device. In this section we will compare some techniques to find the energy centre of the signal and eventually discuss their advantages.

5.1. Time-series simple statistics algorithms for peak detection

Peak Signal Algorithms are simple statistic algorithms for non-normally distributed data series [23] to find the peak signal through threshold criteria statically calculated [23]. The algorithms which identify peaks in a given normally distributed time-series were selected to be applied in a power distribution data, whose peaks indicate high demands, and the highest corresponds to the energy centre. Each different algorithm is based on specific formalization of the notion of a peak according to the characteristics of the optical signal. These algorithms are classified as simple since the signal does not require to be pre-processed to smooth it, neither to be fit to a known function. However, the used algorithm detects all peaks whether strong or not, and to reduce the effects of noise it is required that the signal-to-noise ratio (SNR) should be over a certain threshold [23]:

$$h = \frac{max + abs\_avg)}{2} + K^*abs\_dev \tag{4}$$

$$S_1(k,\ i,\ x_i,\ T) = \frac{max\{x_i - x_{i-1},\ x_i - x_{i-2},\ x_i - x_{i-k}\} + max\{x_i - x_{i+1},\ x_i - x_{i+2},\ x_i - x_{i+k}\}}{2} \tag{5}$$

$$S_2(k,\ i,\ x_i,\ T) = \frac{\frac{x_i - x_{i-1} + x_i - x_{i-2} + \cdots + x_i - x_{i-k}}{k} + \frac{x_i - x_{i+1} + x_i - x_{i+2} + \cdots + x_i - x_{i+k}}{k}}{2} \tag{6}$$

$$S_3(k,\ i,\ x_i,\ T) = \frac{\left(x_i - \frac{x_i - x_{i-1} + x_i - x_{i-2} + \cdots + x_i - x_{i-k}}{k}\right) + \left(x_i - \frac{x_i - x_{i+1} + x_i - x_{i+2} + \cdots + x_i - x_{i+k}}{k}\right)}{2} \tag{7}$$

$$S_4(k,\ i,\ x_i,\ T) = H_w(N(k,\ i,\ T)) - H_w(N'(k,\ i,\ T)) \tag{8}$$

This method consists in define the variables: T = x1, x2, ...,x Nbe a given univariate uniform-ly sampled time-series containing N values (1,2, ...,N). xi be a given it h point in T. k> 0 is a given integer. N+(k,i,T) = <xi+1, xi+2,...,xi+k> the sequence of k right temporal neighbours of xi. N-(k,i,T) = <xi-1, xi-2,...,xi-k> the sequence of k left temporal neighbours of xi.N(k,i,T) = N +(k,i,T) • N-(k,i,T) denote the sequence of 2k points around the it h point (without the ith point itself) in T (• denotes concatenation). N'(k,i,T) = N+(k,i,T) • {xi} • N-(k,i,T). And S be a given peak function, (which is a non-negative real number).S(i, xi, T) with it h element xi of the given time-series T.

A given point xi in T is a peak if S(i, xi, T) >θ, where θ is a user-specified (or suitably calcu-lated) threshold value.

5.2. Calculation of the centroid of the light distribution

This method has been widely used in digital imaging for the location of different image fea-tures with subpixel accuracy. By definition, the centroid of a continuous 1-D light intensity distribution is given by:

$$x = \frac{\int_{-\infty}^{\infty} xf(x)dx}{\int_{-\infty}^{\infty} f(x)dx} \tag{9}$$

where f (x) is the irradiance distribution at the position x on the image, see [24].

In our case the signal geometric centroid is a function of the voltage signal shape generated by the scanner (a plane figure of two dimensional shape X) and is the intersection of all straight lines that divide X into two parts of equal moment about the line [25].

For the geometric centroid computation the "integral plane figures method" will be used to provide two coordinates: $\dot{X}$ which we will assign to the time axis, thus $\dot{X}=T$ and $\dot{Y}$ that we will assign to the voltage axis, thus $\dot{Y}=V$, where we will only take the t coordinate to corre-late the geometric centroid with the position on time (sample number) where the energy centre is located. The signal generated by the optical aperture should be represented by a function:

$$y = f(t) = v(t) \tag{10}$$

In Figure 21 the area under the curve delimited by the function y=v(t) and the lines Aa and Bb define the function integral limits of the plane figure. Selecting the differential area

$$dA = v(t)\, dt \tag{11}$$

The integral limits are on t (a, b). As the differential dt is a rectangle, the geometric centroid is in the half base and half height. As dt tends to zero and the half of it is a very small value, we could consider that half of dt is dt, therefore the next equations are used to calculate the geometric centroid [5]:

$$T = tV = v/2 \tag{12}$$

First Integral, to calculate the complete area under the signal curve.

$$A = \int dA = \int v(t)dt \tag{13}$$

Second Integral, to find the T coordinate that corresponds to the energetic signal centre.

$$TA = \int TdA = \int tv(t)dt \tag{14}$$

Solving for T

$$T = \frac{\int TdA}{A} = \frac{\int tvdt}{A} \tag{15}$$

Third Integral, to find the V coordinate to know which voltage value was present in the energetic signal centre coordinate, if required for future experimentation.

$$VA = \int VdA = \int (v/2)vdt = 1/2\int v^2dt \tag{16}$$

Where

$$V = \frac{\int V da}{A} = \frac{\int (v\,/\,2)vdt}{A} = \frac{1}{2A}\int v^2 dt \qquad (17)$$

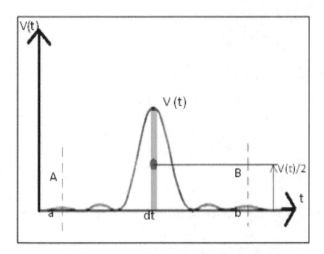

Figure 21. Optical Signal represented as a plane figure.

5.3. Power (energy) spectrum centroid

The Power Spectrum Centroid is a parameter from the spectrum characterization mainly used until now for musical computing due to the spectral centroid corresponding to a timbral feature that describes the brightness of a sound. In this application we will correlate the Power Spectrum Centroid with the Energy Centre of the Signal due to the fact that the Power Spectrum Centroid can be thought as the centre of gravity for the frequency components in a spectrum. The power spectrum is a positive real function of a frequency variable associated with the function of time, which has dimensions of power per hertz (Hz) or energy per hertz, that means the power carried by the wave (signal) per unit frequency.

Power Spectrum Method:

The first step to go from time-series domain to frequency-series domain is to apply the Fourier series, which provides an alternate way of representing data. Instead of representing the signal amplitude as a function of time, Fourier Series represent the signal by how much information (power) is contained at different frequencies and also allow to isolate certain frequency ranges that could be from noise sources, if necessary. Whenever we have a vector of data (finite series)with Matlab we can apply the FFT (Fast Fourier Transform) to convert from time to frequency domain, computing the Discrete Fourier transform (DFT), which is the Fourier application for discrete data and whose non-zero values are finite series.

The second step is to compute the power spectrum, that is to compute the square of the absolute value of the FFT, which result is considered as the power of the signal at each frequency.

$$\Upsilon() = \left| \frac{1}{\sqrt{2\pi}} \sum_{n=-\infty}^{\infty} f_n e^{-i\omega n} \right| 2 = \frac{F(\omega)F^*(\omega)}{2\pi} \qquad (18)$$

Where F(ω) is the discrete-time Fourier Transform of fn.

The third step consists of applying the Power Spectrum Centroid:

$$SC_{Hz} = \frac{\sum_{k=1}^{N-1} k.\, X^d[k]}{\sum_{k=1}^{N-1} X^d[k]} \qquad (19)$$

6. Method and electronic device to locate signal energy centre

The principal focus of this chapter is a method to find the energy centre of the signal generated by optical scanners and to reduce errors in position measurements. The method is based on the assumption that the signal generated by optical scanners for position measurement is a Gaussian-like shape signal, and this signal is processed by means of an electronic circuit.

6.1. Electronic method operating theory

A signal V(t) is obtained from the optical scanning aperture, as shown in Figure 22.

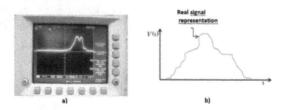

Figure 22. a) Chanel 1 Original signal from apertureb) Original signal representation.

The signal V(t) is amplified through an operational amplifier until saturation to obtain a square signal, this signal can be expressed as:

$$V_s(t) = V_s max \qquad (20)$$

which is a constant for a≤t ≤bas shown in Figure 23.

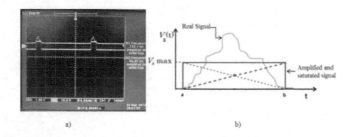

a)   b)

Figure 23. a) Channel 1 Original Signal from aperture. Channel 2 Square Signal. b) Square signal representation.

The signal Vs (t) is integrated with respect to dt in order to get the ramp Vr(t) as shown below in Figure 24.

$$V_r(t) = \int V_s(t)dt \tag{21}$$

then energetic signal centre is located in Vrmax/2= Vsmax /2 as shown in Figure 24.

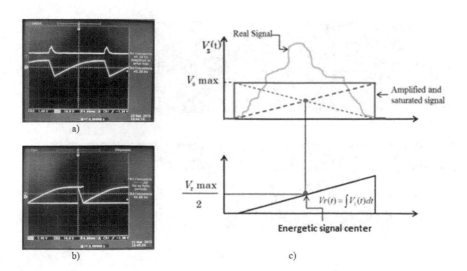

Figure 24. a)Channel 1 original signal from aperture, Channel 2 ramp signal. b) Channel 1 ramp signal, Channel 2 Pulse indicating the energetic signal centre overlapped on ramp signal c) Energetic signal centre search process representation.

All this process is carried out by a circuit similar to the one shown in figure 25.

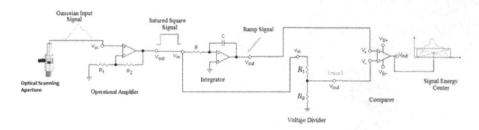

Figure 25. Electronic control circuit representation.

The advantage of this method is that the mathematical processing is performed by using electronic components in real time to get the data vector of the saturated signal which can be handled in Matlab [8].

6.2. Electronic control circuit method experimentation

The first stage of the experimentation started with regular signals simulated by a function generator, the signals utilized were: a rectified sin signal, a rectified square signal, rectified triangle signal and an Airy function, as illustrated in figure 26.

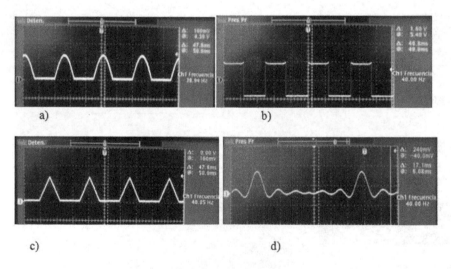

Figure 26. Regular signals simulated by a function generator: a)Rectified Sin signal. b) Square ignal. c) Rectified triangle signal d)Airy function signal.

The second stage consisted of processing each signal by means of the electronic circuit to get the energetic centre, as shown in figure 27.

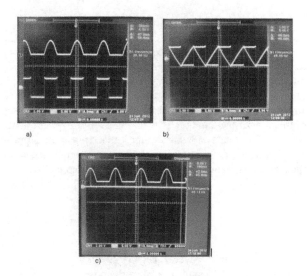

Figure 27. Detailed example. a)Channel 1: original signal from function generator; Channel 2: square signal obtained from saturation. b) Channel 1: saturated signal Channel 2: ramp signal obtained from integration. c) Channel1: original signal from generator; Channel 2: impulse signal overlapped on original signal from generator to indicate the energetic signal centre.

The circuit was tested with different signals obtaining satisfactory results. In figure 28, the results are illustrated with a triangle signal and an airy function signal.

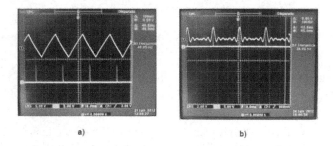

Figure 28. Impulse signal indicating the energetic signal centre overlapped on: a) triangle signal. b)Airy function signal.

Finally, to characterize and increase the accuracy and resolution of signal measurements four methods were selected and compared to obtain the most applicable settingsin order to find the energetic centre of the signal. The given results are shown in [2].

7. Conclusion

A pertinent method to detect the energetic centre of signal generated by optical scanning with a rotating mirror and a simple photodiode was presented. The results of a series of experiments and simulations were used to analyse the performance of the method and the circuit considering regular signals. Consequently, further work is required to reduce problems encountered in processing real signals. Besides, The method can also be used to detect geometrical centre of the light distribution on CCD and PSD, future experiments with this kind of sensors should be considered. The results suggest that the circuit proposed can support different patterns of light distributions. To conclude, it is strongly recommended that this circuit and the photodiode be manufactured in the same integrated circuit.

Author details

Moisés Rivas, Wendy Flores, Javier Rivera, Oleg Sergiyenko,
Daniel Hernández-Balbuena and Alejandro Sánchez-Bueno

Engineering Institute of Autonomous University of Baja California (UABC), Mexico

References

[1] Rivas, L., Moisés, Sergiyenko., Oleg, Tyrsa., Vera, Hernández., & Wilmar, . Optoelectronic method for structural health monitoring. International Journal of Structural Health Monitoring (2010). , 9(1), 105-120.

[2] Flores, F., Wendy, Rivas. L., Moisés, Sergiyenko., Oleg, Y., Rivera, C., & Javier, . Comparison of Signal Peak Detection Algorithms in the Search of the Signal Energy Center for measuring with optical scanning. In: Falcon S. Bertha, Trejo O. René, Gómez E. Raúl: IEEE ROC&C2011:XXII autumn international conference on communications, computer, electronics, automation, robotics and industrial exposition:ROC&C2011, 27 Nov-1 Dec. (2011). Acapulco Gro., México.

[3] Rodríguez Q. Julio C., Sergiyenko Oleg, Tyrsa Vera., Básaca P Luis C., Rivas L. Moisés, Hernández B. Daniel, 3D Body& Medical Scanners`Technologies: Methodology and Spatial Discriminations, in: Srgiyenko Oleg (ed.) Optoelectronic Devices and Proprieties. In-Tech: 2011.p307-322.

[4] Kennedy, William. P. The Basics of triangulation sensors, Cyber Optics Corp. http://archives.sensorsmag.com/articles/0598/tri0598/main.shtml,accessed June 18 (2012).

[5] Vahelal, Ahmedabad. Seminar report 3 of Cryptography: Sensors on 3D Digitization, Hasmukh Goswami College of Engineering, Gujarat, India:. (2010).

[6] M. Bass, Handbook of Optics, Vol. II- Devices, Measurements and Properties, New York: McGRAW-HILL , INC, 1995.

[7] J. P. *. a. C. B. Javier Plaza, Multi-Channel Morphological Profiles for Classification of Hyperspectral Images Using Support Vector Machines, Sensors:2009.p197

[8] Rivas M., Sergiyenko O., Aguirre M, Devia L, Tyrsa V, Rendón I, Spatial data acquisition by laser scanning for robot or SHM task: IEEE-IES: International. Symposium on Industrial Electronics:(ISIE-2008), Cambridge, United Kingdom, 2008, p.1458-1463

[9] Rieke, George. Detection of Light: From the Ultraviolet to the Sub millimeter.Optical detectors.(2002). Publisher: Cambridge University Press. West Nyack, NY, USA. eUSBN: 978113948351. pISBN:9780521816366.

[10] Burr-Brown, Corporation. P. D., S-12, , 199, B., & , O. P. (1994). OPT301. Integrated photodiode and amplifier datasheet.

[11] Everlight, Electronics., Co, , Ltd, D. P., & T-03305 , . (2005). PT334-6C. Technical data sheet 5mm phototransistor T-1 3/4.

[12] Jung, R. Image sensors technology for beam instrumentation, CERN, CH1211 Geneva 23, Switzerland

[13] Albert, J. P., & Theuwissen, Solid. Solid-State Imaging with Charge-Coupled Devices. (1995). Kluwer Academic Publishers.

[14] Texas, Instruments. S. O. C., S0, , & , D. (2003). TC281.X1010-Pixel CCD image sensor datasheet., 1036.

[15] Sebastiano Battiato. Arcangelo Ranieri Bruna. Giuseppe Messina.Image Processing for Embedded Devices. (2010). Bentham Science Publisher. Sharjah, UAE.

[16] ADIMEC. CCD vs. CMOS Image Sensors in Defense Cameras. 2011.http:// info.adimec.com/blogposts/bid/58840/CCD-vs-CMOS-Image-Sensors-in-Defense-Cameras

[17] Boni Fabio, What is the difference between PSD and CCD sensor technology?, FA-SEP.http://www.fasep.it/english/support/tech_talks/tt013_psd%20and%20ccd.aspaccessed 20 Jun (2012).

[18] Williams, R. D., Schaferg, P., Davis, G. K., & Ross, R. A. Accuracy of position detection using a position sensitive detector, "IEEE transactions on instrumentation and measurement. (1998). , 1998(47), 4-914.

[19] Massari, N., Gottardi, M., Gonzo, L., & Simoni, A. High Speed Digital CMOS 2D Optical Position Sensitive Detector in:Castello C., Baschirotto A. (eds.) ESSCIRC 2002.Proceedings of 28th European Solid-State Circuits Conference of IEEE Solid-State Circuits Society: ESSCIRC 2002,September (2002). Firenze, Italy., 24-26.

[20] van Oijen A.M., J. Köhler, Schmidt J., Müller M., Brakenhoff G.J.3-Dimensional su-
      per-resolution by spectrally selective imaging Chemical Physics Letters, 1998;
      292(1998)183-187

[21] Rivas L. Moisés, Sergiyenko Oleg, and Tyrsa Vera. Machine Vision: Approaches and
      Limitations. In: Xiong Zhihui. (ed.) Computer Vision. InTech; 2008.p396-428.

[22] Sergiyenko Oleg ,Tyrsa Vira, Basaca P. Luis., Rodríguez Q. Julio C., Hernandez W.,
      Nieto H.Juan I. , Rivas L M. Electromechanical 3D optoelectronic scanners: resolution
      constraints and possible ways of its improvement. In: Seriyenko (ed.) Optoelectronic
      Devices and Properties. InTech: 2011p549-582.

[23] Girish Keshav Palshikar, Simple Algorithms for Peak Detection in Time-Series, Tata
      Research Development and Design Centre (TRDDC), 54B Hadapsar Industrial Es-
      tate.Pune 411013,India.

[24] Stanton, R., Alexander, J. W., Dennison, E. W., Glavich, T. A., & Hovland, L. F. Opti-
      cal tracing using charge-coupled devices. (1987). Opt. Eng. , 26(9), 930-938.

[25] Larry Herrera, Magister in Educational Planning, Oriente University of Bolívar,
      http://www.udobasico.net/mecanica/CLASES%20centroide%204.html (accessed 10
      Jun 2012).

Chapter 7

III-V Multi-Junction Solar Cells

Gui jiang Lin, Jingfeng Bi, Minghui Song,
Jianqing Liu, Weiping Xiong and Meichun Huang

Additional information is available at the end of the chapter

1. Introduction

Photovoltaic is accepted as a promising technology that directly takes advantage of our planet's ultimate source of power, the sun. When exposed to light, solar cells are capable of producing electricity without any harmful effect to the environment or devices. Therefore, they can generate power for many years (at least 20 years) while requiring only minimal maintenance and operational costs. Currently the wide-spread use of photovoltaic over other energy sources is impeded by the relatively high cost and low efficiency of solar cells [1].

III-V multi-junction solar cells, as a new technology, offer extremely high efficiencies compared with traditional solar cells made of a single layer of semiconductor material [2]. The strong demand for higher efficiency photovoltaic has recently attracted considerable interest in multi-junction solar cells based on III-V semiconductors [3]. Depending on a particular technology, multi-junction solar cells are capable of generating approximately twice as much power under the same conditions as traditional solar cells made of silicon. Unfortunately, multi-junction solar cells are very expensive, so they are mainly used in high performance applications such as satellites at present. However, in our opinion, with the concentrator technology, the tandem cell will play a role in the future energy market. The state-of-the-art high efficiency III-V solar cells utilize a triple junction structure which consists of the Ge bottom sub-cell (0.67 eV) formed on the Ge substrate homogeneously, the $Ga_{0.99}In_{0.01}As$ middle sub-cell (1.36 eV), and the lattice matched (LM) $Ga_{0.5}In_{0.5}P$ top sub-cell (1.86 eV) [4,5]. It has reached conversion efficiencies up to 40% at concentrations of hundreds of suns under the AM1.5D low aerosol optical depth (AOD) spectrum [4]. The GaInP/GaInAs/Ge triple-junction cells have also been demonstrated with efficiencies up to 30% under one-sun AM0 spectrum for space applications. Multi-junction solar cells based on III-V materials have achieved the highest efficiencies of any present photovoltaic devices. Additionally, these devices are the only solar cells currently available with efficiencies above 30%. The high efficiency is due to the reduction of thermalization and transmission losses in solar cells when the number of p-n junctions is increased.

Future terrestrial cells will likely feature four or more junctions with a performance potential capable of reaching over 45% efficiency at concentration of hundreds of suns. The 4-, 5-, or 6-junction solar cells with concentrator trade lower current densities for higher voltage and divide the solar spectrum more efficiently. The lower current densities in these cells can significantly reduce the resistive power loss (I^2R) at high concentrations of suns when compared with the 3-junction cell [6].

High-efficiency GaInP/GaAs/InGaAs triple-junction solar cells grown inversely with a metamorphic bottom junction could be achieved by replacing the bottom Ge sub-cell with 1 eV energy gap material. $In_{0.3}Ga_{0.7}As$ is the promised candidate, if without the lattices mis-match (LMM, around 2%) with the other two sub-cells. Therefore, the LM top and middle sub-cells were grown first, and the graded buffers were employed between middle and bottom cells to overcome the mismatch and to prevent the threading dislocations. The substrate was removed for the reusing. This inverted metamorphic, monolithic triple junction solar cells could be obtained with at least 2% higher efficiency than the traditional one theoretically [7].

A metamorphic $Ga_{0.35}In_{0.65}P/Ga_{0.83}In_{0.17}As/Ge$ triple-junction solar cell is studied to provide current-matching of all three sub-cells and thus give a device structure with virtually ideal energy gap combination. It is demonstrated that the key for the realization of this device is the improvement of material quality of the lattice-mismatched layers as well as the development of a highly relaxed $Ga_{1-y}In_yAs$ buffer structure between the Ge substrate and the middle cell. This allows the metamorphic growth with low dislocation densities below 10^6 cm^{-2}. The performance of the device has been demonstrated by a conversion efficiency of 41.1% at 454 suns AM1.5D [8].

In this chapter, the theoretical and experimental investigation of the most sophisticated, industrialized and commercialized GaInP/GaInAs/Ge triple junction solar cell was extensively described. Accelerated aging tests of the high concentration multi-junction solar cells and discussions on outdoor power plant performances were also presented.

2. Theoretical study on optimization of high efficiency multi-junction solar cells

In designing GaInP/GaInAs/Ge triple-junction cells, the principles for maximising cell efficiency are: (1) increasing the amount of light collected by each cell that is turned into carriers, (2) increasing the collection of light-generated carriers by each p-n junction, (3) minimising the forward bias dark current, and (4) photocurrents matching among sub-cells.

In practice, basic designs for these solar cells involve various doping concentrations and layer thicknesses for the window, emitter, base, and back surface field (BSF) regions in each sub-cell. In order to optimize the designs, a rigorous model including optical and electrical modules was developed to analyse the bulk parameters effect on the external quantum efficiency, photocurrent and photovoltage of the GaInP/GaInAs/Ge multi-junction solar cells.

2.1. Theoretical approach

We present here a brief description of the equations used in our model. Thorough treatments of photovoltaic devices can be found elsewhere [9]. A schematic of a typical lattice-matched GaInP/GaInAs/Ge solar cell is shown in Figure1. It consists of an n/p GaInP junction on top of an n/p GaInAs junction which lies on an n/p Ge junction. The triple junction cells are series connected by two p++/n++ tunnel junctions.

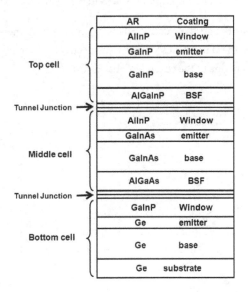

Figure 1. Solar cell structure used for simulation.

The solar spectrum, striking the front of the cell, includes ultraviolet, visible, and infrared lights. The absorption coefficient for short-wavelength light is quite large, and most of the blue light is absorbed very close to the front of the cell for generating photo carriers. Light with energy slightly larger than the energy gap is weakly absorbed throughout the cell. Light with energy less than the energy gap passes through the front cell and is absorbed in the next one. The photo carriers generated by the short-wavelength light diffuse inside the cell until they are either collected at the p-n junction or recombined with a majority carrier in bulk or at interface. The efficiency of the solar cell increase when all the photo carriers are collected at the junction instead of recombining elsewhere. Thus, recombination is at the front and back of the cell effects on the efficiency of the cell.

At the first level approximation, multi-junction cells behave like homo-junction cells in series, so their open circuit voltage is the summation of the voltages of the sub-cells, while their short circuit current is that of the sub-cell with the smallest current. Hence, the performance of a multi-junction cell can be obtained from the performance of each sub-cell, evaluated independently. The load current density J is represented by the superposition of two diode currents and the photo-generated current,

$$J = J_{ph} - J_{01}(e^{qV/kT} - 1) - J_{02}(e^{qV/2kT} - 1) \tag{1}$$

Where J_{ph} is the photocurrent density, J_{01} the ideal dark saturation current component and J_{02} the space charge non-ideal dark saturation current component.

The photocurrent density and dark current density are given by the sum of the photocurrents and the sum of the dark current density, respectively, generated in the emitter, the base and the depleted region of the cell. [9] We have

$$J_{ph} = J_{emitter} + J_{base} + J_{depleted} \tag{2}$$

$$J_{emitter} = \frac{qF(1-R)\alpha L_p}{(\alpha L_p)^2 - 1} \left\{ \frac{S_p L_p}{D_p} + \alpha L_p - e^{-\alpha(d_e - W_n)} \right. \\ \left. \left(\frac{\frac{S_p L_p}{D_p}\cosh[(d_e - W_n)/L_p] + \sinh[(d_e - W_n)/L_p]}{\frac{S_p L_p}{D_p}\sinh[(d_e - W_n)/L_p] + \cosh[(d_e - W_n)/L_p]} \right) \right. \\ \left. - \alpha L_p e^{-\alpha(d_e - W_n)} \right\} \tag{3}$$

$$J_{base} = \frac{qF(1-R)\alpha L_n}{(\alpha L_n)^2 - 1} e^{-\alpha(d_b - W_n + W)} \left\{ \alpha L_n - \frac{\frac{S_n L_n}{D_n}\left(\cosh[(d_e - W_n)/L_n] - e^{-\alpha(d_b - W_p)}\right)}{\frac{S_n L_n}{D_n}\sinh[(d_b - W_p)/L_n] + \cosh[(d_b - W_p)/L_n]} \right\} \tag{4}$$

$$J_{depleted} = qF(1-R)e^{-\alpha(d_e - W_n)}\left(1 - e^{-\alpha W}\right) \tag{5}$$

$$J_{01} = J_{01,emitter} + J_{01,base} \tag{6}$$

$$J_{01,emitter} = q \frac{n_i^2}{N_D} \frac{D_p}{L_p} \left\{ \frac{S_p L_p / D_p \cosh[(d_e - W_n)/L_p] + \sinh[(d_e - W_n)/L_p]}{S_p L_p / D_p \sinh[(d_e - W_n)/L_p] + \cosh[(d_e - W_n)/L_p]} \right\} \quad (7)$$

$$J_{01,base} = q \frac{n_i^2}{N_A} \frac{D_n}{L_n} \left\{ \frac{S_n L_n / D_n \cosh[(d_b - W_p)/L_n] + \sinh[(d_b - W_p)/L_n]}{S_n L_n / D_n \sinh[(d_b - W_p)/L_n] + \cosh[(d_b - W_p)/L_n]} \right\} \quad (8)$$

$$J_{02} = \frac{W n_i}{2(V_d - V)\tau} \quad (9)$$

Where q is electron charge, F the incident photon flux, α is an optical absorption coefficient and R is the reflectance of the anti-reflective coating. n_i is the intrinsic carrier concentration, N_A and N_D are the concentrations of acceptors and donors. d_e is the emitter thickness, d_b the base thickness, L_p the hole diffusion length in the emitter, L_n the electron diffusion length in the base, S_p the hole surface recombination velocity in the emitter, S_n the electron surface recombination velocity in the base, D_p the hole diffusion coefficient in the emitter, D_n the electron diffusion coefficient in the base, and τ is the non-radiative carrier lifetime. T_F is the transmission of incident photon flux into the sub-cell under consideration.

The build-in voltage V_d of the junction, the thickness of the depleted layer in the emitter W_n, the thickness of the depleted layer in the base W_p, and the total depleted zone thickness W, are given by [10],

$$V_d = kT \log(\frac{N_D N_A}{n_i^2}) \quad (10)$$

$$W = \sqrt{2\varepsilon \frac{N_D + N_A}{N_D N_A}(V_d - V - 2kT)} \quad (11)$$

$$W_n = W / (1 + N_D / N_A) \quad (12)$$

$$W_p = W - W_n \quad (13)$$

Where k is the Boltzmann constant, ε the dielectric constant and T the temperature (T =25 °C was used in this paper). It is important to note that F and α depend on the wavelength, whereas D_p, D_n, L_p, L_n and τ depend on the doping concentration [11].

The optical model proposed in this paper is based on the transfer matrix formalism. It allows calculating the incident optical spectrum on each sub-cell from the solar spectrum. Each layer of the multi-junction is described by a transfer matrix M which is defined by

$$M = \begin{pmatrix} M_{0,0} M_{0,1} \\ M_{1,0} M_{1,1} \end{pmatrix} = \begin{vmatrix} \cos\left(d\frac{2\pi(n-i\lambda\alpha/4\pi)}{\lambda}\right) & i\frac{\sin\left(d\frac{2\pi(n-i\lambda\alpha/4\pi)}{\lambda}\right)}{(n-i\lambda\alpha/4\pi)} \\ (n-i\lambda\alpha/4\pi)\sin\left(d\frac{2\pi(n-i\lambda\alpha/4\pi)}{\lambda}\right) & \cos\left(d\frac{2\pi(n-i\lambda\alpha/4\pi)}{\lambda}\right) \end{vmatrix} \tag{14}$$

Where n and d are the refraction index and the thickness of the layer, respectively. The transmission coefficient T_M [12] of the layer is then given by

$$T_M = \frac{4n_0^2}{(n_0 M_{0,0} + n_0 n_s M_{0,1} + M_{1,0} + n_s M_{1,1})^2} \tag{15}$$

Where n_0 is the superstrate refraction index and n_s is the substrate refraction index of the sub-cell. The $M_{i,j}$ coefficients refer to the matrix transfer elements. Thus, it is possible to determine the incident spectrum on each sub-cell. The incident photon flux in GaInP, GaInAs and Ge sub-cells are given by

$$F_{GaInP} = T_{ARC} F_{solar} \tag{16}$$

$$F_{GaInAs} = T_{ARC} T_{GaInP} F_{solar} \tag{17}$$

$$F_{Ge} = T_{GaInAs} T_{GaInP} T_{ARC} F_{solar} \tag{18}$$

where F_{solar} is the incident photon flux, T_{ARC}, T_{GaInP} and T_{GaInAs} are the transmission coefficient of the anti-reflective coating, the GaInP sub-cell and the GaInAs sub-cell, respectively. This model includes optical and electrical modules. Thus, it allows the calculation of the quantity of photons arriving at each junction from the solar spectrum. Then, the electrical model calculates, via interface recombination velocity, the photocurrents in the space charge region, the emitter and the base for each junction.

2.2. Solar cell structures and parameters

To calculate the power production of the GaInP/GaInAs/Ge triple-junction cells for space applications, the incident photon flux F_{solar} was taken from a newly proposed reference air mass zero (AM0) spectra (ASTM E-490). The integration of ASTM E490 AM0 solar spectral irradiance has been made to conform to the value of the solar constant accepted by the space community, which is 1366.1 W/m². The transmission coefficient of the anti-reflective coating T_{ARC} was set to be a constant of 98%, while the transmission coefficients of the GaInP sub-cell and the GaInAs sub-cell are calculated according to Eqs.14 and 15, which have wavelength dependence.

Parameter	Ge	GaInAs	GaInP
D_n(cm²/s)	22.86	140.02	29.39
D_p(cm²/s)	10.71	4.02	1.03
L_n(cm)	5.3×10^{-3}	9.7×10^{-4}	6.3×10^{-4}
L_p(cm)	8.8×10^{-4}	7.3×10^{-5}	3.7×10^{-5}
τ (s)	8.9×10^{-7}	8.9×10^{-9}	4.2×10^{-9}

Table 1. Material parameters used for calculation in this paper.

As shown in Figure 1, typical two-terminal triple-junction cells for space application with a Ge bottom cell, a GaInAs middle cell and a GaInP top cell with energy gaps of 0.661, 1.405 and 1.85 eV, respectively. The Ge cell is built on the p-type initial substrate; therefore, the Ge base is about 150 micrometers thick, with doping concentration of about 6×10^{17} cm⁻³; the Ge emitter is about 0.3 micrometers thick, with an n-type doping concentration of about 1×10^{19} cm⁻³. The emitters for the other two cells are 0.1 micrometers thick with doping concentration of about 1×10^{18} cm⁻³. Since the AM0 spectrum contains relatively more high-energy photons with energy greater than the GaInP top cell's energy gap, triple-junction cell with a very thick top cell will generally be photocurrent limited by the middle (GaInAs) cell. Therefore, the middle cell thickness was set to be thick enough (3.6 micrometers in this paper) with the doping concentration of about 2×10^{17} cm⁻³, and the optimal top cell thickness was suggested to be about 0.52 micrometers with doping concentration of about 1×10^{17} cm⁻³.

The absorption coefficient of the GaInP can be fitted by

$$\alpha_{GaInP} = 5.5\sqrt{(E - E_g)} + 1.5\sqrt{(E - E_g - 1)} \tag{19}$$

The absorption coefficient of the GaInAs (with In content of about 0.01) can be fitted by

$$\alpha_{GaInAs} = 3.3\sqrt{(E - E_g)} \tag{20}$$

The direct gap absorption spectra of the bulk Ge was used for calculation

$$\alpha_{Ge} = 1.9\sqrt{(E - E_g^\Gamma)} \big/ E \tag{21}$$

Where E is the photon energy and E_g the fundamental energy gap, both in eV, and α · in 1/micrometers.

The diffusion length, the diffusion coefficient and the nonradiative carrier lifetime are calculated as a function of the doping concentration. The material parameters used for calculation are summarized in table 1.

2.3. The effect of the interface recombination on the performance of GaInP/GaInAs/Ge tandem solar cell

To have an analytical analysis, recombination velocity at only one interface among six interfaces is assumed to have a non-zero value, which is 1×10^6 cm/s. Figure2~4 shows the total external quantum efficiencies η and the integrated photocurrent density J_{ph} of the three sub-cells, calculated from Eqs. 2-5 with the constant parameters in table 1 and with varying values of S_p and S_n. The external quantum efficiency η, defined as the probability of collecting a photo carrier for each photon, is a function of wavelength, λ, because of the λ-dependence of the absorption coefficient, α. The photocurrent density J_{ph} is obtained from the integral of the product of the η with the spectrum of interest. For large absorption coefficients, a high S_p causes dramatic decrease in the blue response as shown in Figure 2 (a), Figure 3 (a) and Figure 4 (a).

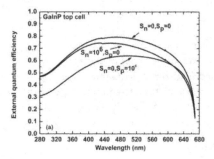
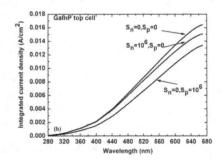

Figure 2. a) External quantum efficiency, and (b) integrated photocurrent density of the top GaInP cell for various interface recombination velocities.

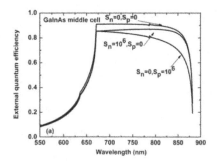
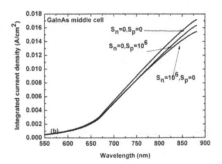

Figure 3. a) External quantum efficiency, and (b) integrated photocurrent density of the middle GaInAs cell for various interface recombination velocities.

However, a high S_p also causes a reduction in the red response. In contrast, high S_n causes a reduction only in the red response (Figure 2 (a), Figure 3(a)), with almost no measurable effect in the blue response for a thick cell as shown in Figure 4 (a).

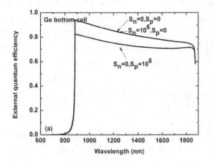 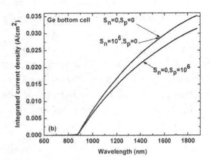

Figure 4. a) External quantum efficiency, and (b) integrated photocurrent density of the middle GaInAs cell for various interface recombination velocities.

Once the photocurrents of the three sub-cells are calculated, the short circuit current of the tandem cell is set to be the smallest of these three photocurrents. The open-circuit voltage is set to be the voltage at which the magnitude of the dark currents equals the photocurrents. The corresponding I-V characteristics of the tandem cell are plotted in Figure 5. Among all the interfaces, recombination at the top cell emitter surface is most detrimental due to the considerable drop of the cell short circuit current and to a less extent to the associated reduction in the cell voltage. While recombination effect at back interface of the bottom cell can be almost negligible because the base layer of the cell is thick enough.

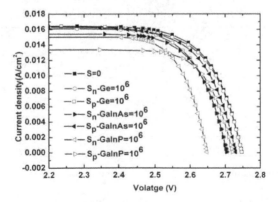

Figure 5. I-V characteristics of the GaInP/ GaInAs/ Ge tandem cell under AM0 with a recombination velocity at the indicated interface and zero elsewhere.

2.4. Optimization of high efficiency GaInP/GaInAs/Ge multi-junction solar cells

Lattice-matched GaInP/GaInAs/Ge triple-junction cells under investigation include a Ge bottom cell, a GaInAs middle cell and a GaInP top cell with energy gaps of 0.661, 1.405 and 1.85 eV, respectively. The Ge cell is built on the p-type initial substrate; the Ge base is 150 micrometers thick with doping concentration of 6×10^{17} cm^{-3}, and the Ge emitter is 300 nm thick with an n-type doping concentration of 1×10^{19} cm^{-3}. The middle cell's base is set to be thick enough (3.6 micrometers in this paper) with doping concentration of 2×10^{17} cm^{-3}, and its emitter is 100 nm thick with doping concentration of 1×10^{18} cm^{-3}. The incident photon flux is taken from a newly proposed reference air mass zero (AM0) spectra (ASTM E-490). The anti-reflective coating used in simulation includes a 30 nm AlInP top window layer; ARC composed of 52 nm ZnS and 90 nm MgF2.

It is at first assumed that recombination velocity for a top cell back surface is 1.3×10^5 cm/s, a middle cell back surface 105 cm/s and a top cell emitter surface 5.15×10^4 cm/s, while recombination velocities at the other three interfaces are zero. Then, the optimal top cell thickness and dopant profiles were obtained to meet high efficiency.

top cell base thickness d-base (nm)	open-circuit voltage Voc (V)	short-circuit current Jsc (A/cm²)	fill factor	tandem cell efficiency
d-base=400 nm	2.6660	0.01720	90.56%	30.40%
d-base=450 nm	2.6664	0.01768	89.82%	30.99%
d-base=500 nm	2.6667	0.01812	88.40%	31.27%
d-base=55 0nm	2.6669	0.01777	89.82%	31.05%

Table 2. Figure-of-merits of the tandem cell for various top cell base thickness.

top cell base doping concentration (1/cm3)	open-circuit voltage Voc (V)	short-circuit current Jsc (A/cm²)	fill factor	tandem cell efficiency
NA-base =1x1016	2.6095	0.01816	88.62%	30.74%
NA-base =5x1016	2.6503	0.01814	89.15%	31.32%
NA-base =1x1017	2.6667	0.01812	88.40%	31.27%
NA-base =5x1017	2.6981	0.01796	88.24%	31.19%

Table 3. Figure-of-merits of the tandem cell for various top cell base doping concentration.

Table 2 presents the Figure-of-merits of the tandem cell for various top cell base thicknesses with doping concentration of 1×10^{17} cm-3, when the top cell emitter thickness is set to 100 nm with doping concentration of 1×10^{18} cm^{-3}. Table 3 presents the Figure-of-merits of the tandem cell for various top cell bases doping concentrations with thickness of 500 nm, when the top cell emitter thickness is set to 100 nm with doping concentration of 1×10^{18} cm^{-3}. It is found that photocurrents strongly depend on top cell thickness, since the AM0 spectrum contains relatively more high-energy photons with energy greater than the GaInP top cell's energy gap, and photocurrents of triple-junction cells with a very thick top cell will generally be limited by the middle (GaInAs) cell. The tandem cell efficiency reaches the largest value (31.27%) with the top cell base thickness of 500 nm, because the photocurrents of the top and middle cells almost match each other. Table 3 shows that higher doping concentration at the top cell base leads to a considerable increase of the cell voltage and a less drop of cell photocurrent. It can be deduced from table 3 that doping concentration at the top cell base should be optimized between 5×10^{16} and 1×10^{17} cm^{-3} to obtain higher efficiency. In order to realize the values of the Figure-of-merits shown in table 2 and table 3, the external quantum efficiency of the top cell for various top cell base thicknesses and top cell base doping concentrations are presented in Figure6 (a) and Figure6 (b), respectively. It is found that the external quantum efficiency of the top cell increases with the increasing top cell base thickness (Figure6 (a)), while at short wavelengths, the efficiency increases with the increasing top cell base doping concentration, at large wavelengths, decreases (Figure6 (b)).

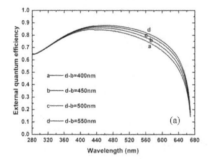
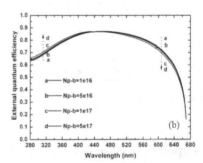

Figure 6. External quantum efficiency of the top cell for various top cell base thickness (a), and top cell base doping concentration (b).

top cell emitter thickness d- emitter (nm)	open-circuit voltage Voc (V)	short-circuit current Jsc (A/cm²)	fill factor	tandem cell efficiency
d-emitter =50 nm	2.6680	0.01848	88.62%	31.98%
d-emitter =100 nm	2.6667	0.01812	88.40%	31.27%
d-emitter =150 nm	2.6710	0.01737	89.09%	30.25%
d-emitter =200 nm	2.6707	0.01644	89.84%	28.87%

Table 4. Figure-of-merits of the tandem cell for various top cell emitter thickness.

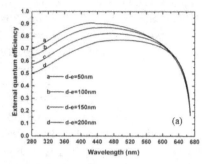
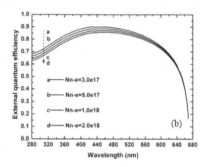

Figure 7. The external quantum efficiency of the top cell for various top cell emitter thickness (a), and top cell emitter doping concentration (b).

top cell emitter doping concentration (1/cm³)	open-circuit voltage Voc (V)	short-circuit current Jsc (A/cm²)	fill factor	tandem cell efficiency
ND -emitter =3x1017	2.6673	0.01816	89.30%	31.66%
ND -emitter =5x1017	2.6674	0.01816	88.37%	31.51%
N D-emitter =1x1018	2.6667	0.01812	88.40%	31.27%
ND -emitter =2x1018	2.6652	0.01786	88.74%	30.90%

Table 5. Figure-of-merits of the tandem cell for various top cell emitters doping concentration.

3. Experimental procedure, results and discussions

3.1. The preparation of the triple junction GaInP/GaInAs/Ge epitaxial wafers

The $Ga_{0.49}In_{0.51}P/Ga_{0.99}In_{0.01}As/Ge$ multi-junction solar cells were grown by the Veeco E475 MOCVD system on 6° off cut Germanium substrate. Standard growth conditions used were with growth pressure of 40 Torr, and rotation rate of 500 rpm. The precursors include trimethylindium (TMIn), trimethylgallium (TMGa), trimethylaluminium (TMAl), arsine, phosphine and diethyl-tellurium (DETe), diethyl-zink (DEZn). Top and middle sub-cells include the following layers: back-surface field (BSF) layer, base, emitter and window. The Ge-sub-cell consists of a base (substrate), a diffused emitter and a window. Sub-cells are connected in series by tunnel diodes, which in turn include highly doped thin (10–20 nm) layers. The growth temperature of 650 °C was applied to the layers consisting of the $Ga_{0.99}In_{0.01}As$ buffer, middle cell layers, top cell layers and GaAs cap. AlGaAs was used as to the middle and top cell BSF, and AlInP as the window layer of InGaAs middle cell and GaInP top cell.

The Ge sub-cell is an important part of the structure of this cell, contributing 10% or more of the total cell efficiency [13]. The Ge junction is formed during III - V /Ge interface epitaxy. Group V elements such as P and As are n-type dopants in Ge, so the emitter of Ge junction was formed by diffusion of V elements during the deposition of III - V epilayers. In addition, the structure of Ge is different from the III - V materials such as GaAs and GaInP, the connection between Ge substrate and buffer layer or initial layer is important to the growth quality on buffer layer and the performance of Ge sub-cell. In this chapter, based on plenty of experiments, GaInP is selected as a suitable buffer material to be grown between the substrate and the active region of the device. Several researches on III - V materials grown on p-doped Ge substrate have indicated that the bottom Ge cell efficiency decreases as the thickness of the emitter increases, mainly owing to the lowering of the short circuit current. For this reason, GaInP is an optimized option with smaller diffusion length than GaAs. In addition, GaInP is also an appropriate material for the window layer of Ge junction.

The electrochemical capacitance-voltage results of GaInP initial layer grown on Ge indicate that the diffusion length of P is about 200 nm, when a thin Ge emitter for excellent performance of Ge sub-cell is fabricated. In the past, GaAs was employed as the middle cell material, and the 0.08% lattice-mismatch between GaAs and Ge was thought to be negligibly. To obtain enough current matched to the top cell, the middle cell was often designed to be 3~4 micrometers thick, but misfit-dislocations were generated in thick GaAs layers and deteriorated cell performance [5]. By adding about 1% indium into the GaAs cell layers, all cell layers are lattice-matched precisely to the Ge substrate. Application of InGaAs middle cell to lattice-match Ge substrates has demonstrated to be able to increase open-circuit voltage (Voc) due to lattice-matching and short-circuit current density (Jsc) due to the decrease of the energy gap in the middle cell.

The $Ga_{0.49}In_{0.51}P/Ga_{0.99}In_{0.01}As/Ge$ multi-junction solar cells for terrestrial concentrator application operate at high current densities higher than $10A/cm^2$. This brings specific challenges to the tunnel diode structures that are used for the series connection of the sub-cells. So the tunnel junction (TJ) growth is one of the most important issues affecting multi-junction solar cell performance. The problems of TJ growth are related to obtaining transparent and uniformly highly doped layer without any degradation of surface morphology [14]. The thickness of each side of the TJ junction has to be in the order of tens of nanometres, while the required doping has to be around 10^{19}~10^{20} cm^{-3}. The reaching of the high doping level requires very different growth temperatures, in order to obtain an abrupt doping profile. In this experiment, the growth of tunnel junction was carried out at temperature of 600 °C which is about 50 °C lower than the growth temperature of other layers. DETe and CCl4 were used as N-type dopant and P-type dopant respectively to fabricate small thickness, high doping AlGaAs/GaAs tunnel junctions.

GaInP lattice-matched to GaAs exhibits anomalous changes in the energy gap, depending on the growth conditions and the substrate misorientation [15]. These changes are the results of the spontaneous ordering during the growth of the cation-site elements (Ga and In) in planes parallel to the (111). One of changes is a lowering of the energy gap of the material, whose exact value depends on the degree of ordering. It appears to be the 100 meV reductions. The

$Ga_{0.49}In_{0.51}P/Ga_{0.99}In_{0.01}As/Ge$ multi-junction solar cells' performance depends on the energy gap of the GaInP top cell. The theoretical calculations for this combination of materials indicate that, to achieve maximum efficiencies, the energy gap of the GaInP top cell should be about 1.89 eV. The GaInP should be completely disordered. However, the MOCVD growth conditions that produce such a material have deleterious effects on the growth quality of GaInP, which determines the performance of the solar cell. To sum up, the growth of high quality GaInP with a maximizing degree of disorder is important for super high efficiency multi-junction solar cells. To fulfil this purpose, precise controls of the growth conditions including the growth temperature, growth rate and V/III ratio were carried out in our experiments. Based on the experimental results and theoretical calculations, the growth of GaInP was carried out at 640 °C, V/III ratio about 40, and growth rates of 0.6 nm/s.

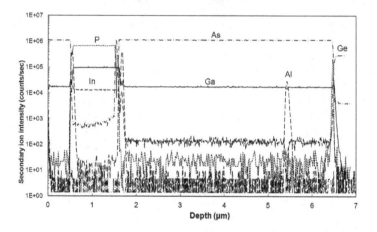

Figure 8. The SIMS spectrum of the $Ga_{0.49}In_{0.51}P/Ga_{0.99}In_{0.01}As/Ge$ multi-junction solar cells.

The SIMS spectrum of the Ga0.49In0.51P/Ga0.99In0.01As/Ge multi-junction solar cells is calibrated and shown in Figure 8. With the elemental depth profile, we can clearly identify the cell structure and the doping level and the thickness of different functional layers.

3.2. The chip processing procedure and optimization

3.2.1. The process procedure of the multi-junction solar cell

The process designed for the concentrator multi-junction solar cells is as follows: the different electrode patterns on the front and back surfaces of the GaInP/ GaInAs/Ge epitaxial wafer are formed first, and then the wafer will be separated into independent cell chips by the methods of chemical etch and/or physical wheel-cutting. Figure 9 shows photos of the GaInP/ GaInAs/Ge epitaxial wafers and chips on wafer process stage. Figure 10 shows the principal process flow. The monolithic device structures of three sub-cells are grown on the

Ge substrate. The graphical front electrode (negative electrode) and the non-graphical back electrode (positive electrode) will then be deposited on the both surfaces of the epitaxial wafers with a series steps of lithography, electrode deposition, metal alloy, cap layer etching, AR coating and so on.

(a)   (b)

Figure 9. GaInP/ GaInAs/Ge epitaxial wafers (a), chips on wafers (b).

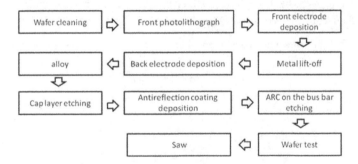

Figure 10. The principal chip process flow.

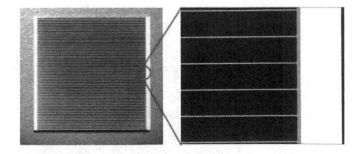

Figure 11. The detail of the graphical front electrode.

Figure11 shows the isolated solar cell and details of the front electrode. Two busbars locate at the edge of the solar cell chip with some parallel gridlines between them. The multi-layer metal structure mainly includes ohmic contact layer, adhesion or barrier layer, conductive layer, and protective layer. The ohmic contact layer will faintly diffuse to the cap layer of the epitaxial wafer after an anneal process, which can decrease the contact resistance between the electrodes and the cap layer. When the photocurrents are generated in the cells, the grid-lines will collect and then transfer the currents to the busbars. Finally the golden wires bonded on the busbars will export the currents to the external circuitry.

3.2.2. Research on the process technology optimization

3.2.2.1. Grid line design

Series resistance (r_S) is the main limiting factor to achieve a high performance for a multi-junction solar cell working under hundreds of suns concentration. Due to the complex con-stituent elements of series resistance, several aspects of the design and manufacture of the solar cells must be considered carefully. The gridline geometries and the metal structure of the triple junction solar cell are the most important factors to reduce the r_S . The main con-cerns are as follows: What is the best r_S value? Which steps during the whole manufacture process will affect the r_S value mostly?

It is known that there are many constituent elements contributed to the series resistance,

$$r_S = r_L + r_V + r_G + r_{FC} + r_{BC} \tag{22}$$

$$\frac{1}{r_L} = \frac{1}{r_E} + \frac{1}{r_W} \tag{23}$$

$$r_V = r_B + r_{Su} \tag{24}$$

where r_L is the resistance of the lateral current in the semiconductor structure, r_V is the re-sistance of the vertical current, r_G is the contribution of gridline, r_{BC} is the resistance of other symbols, and r_{FC} is resistance of the front contact. r_E and r_W are the contribution of emitter layer and window layer to the r_{El} espectively. r_B and r_{Su} are the contribution of base layer and substrate to the r_V respectively.

Traditionally the front metal grid of concentrator solar cells has been thickened up to 5~7 micrometers by electroplating. The higher ratio of the thickness to the width of the grid lines results in larger profile area. On the other hand, the resistance of the gridline metal (r_G) affects the series resistance of the solar cells greatly.

Due to the low thickness (100~150 nm) of the contact semiconductor layer, the lateral resist-ance in the semiconductor (r_L) also plays an important role in the constituent elements of series resistance. An effective method to reduce the r_L , is to decrease the space between the neighbour gridlines, which can be described as the shadowing factor, Fs: a ratio of Area covered by metal to total area. As the shadowing factor increases, the area of the front contact will increase and the series resistance components related to the front contact, r_{FC} , decreases. The lateral current can be easier to collect through a shorter distance. Unfortunately, the shadow of the front grid line increases as the shadowing factor increasing, resulting in a reduction of the Isc of the solar cell. Therefore, the balance between the lower lateral resistance and higher Isc related to the shadowing factor should be considered carefully.

Experimental verification is carried out to obtain the optimum front grid design for 1000 suns concentration GaInP/GaInAs/Ge multi-junction solar cells. Typical values of the front contact resistance (r_{FC}) and the thickness of the grid line metal are 5×10^{-5} $\Omega.cm^2$ and 7 micrometers, respectively. The front contact metal sheet resistance, r_{Msheet} , ranges from 3 to 5×10^{-6} $\Omega.cm^2$. The space between the neighbour gridlines are in the range of 45 ~167 micrometers, respectively. All the Fs and WL values are referred to fingers of 7 micrometers thick in a 10×10 mm2 sized solar cell. It must be pointed that there is no antireflection coating on the surface of all the solar cells discussed here.

Figure12 (a) shows the short-circuit current (Isc), fill factor (FF) and efficiency (Eff) as a function of the shadowing factor. It is evident that with the wider space, the higher Isc can be obtained, because more light can be absorbed by the solar cell. The FF increases obviously as the space increases. Therefore, we can draw a conclusion that the optimal front gridline design should result in higher Isc and FF. As shown in Figure 12 (b), the highest Isc×FF is found with an Fs of 5%, and the corresponding efficiency of 29.8% is also the highest one.

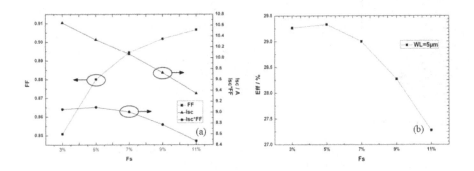

Figure 12. The short-circuit current (Isc), fill factor (FF) and efficiency (Eff) as a function of the shadowing factor.

3.2.2.2. Antireflection Coating

Due to the high refractive indices of semiconductors, high reflection losses must be minimized by antireflection coatings (ARC) for GaInP/GaInAs/Ge multi-junction solar cells. This presents several challenges for the ARC design. Firstly, the wide wavelength range of sunlight requires an optimization of extreme broadband design and limits the material choice to those with little or no absorption over the required wavelength range. For high concentrator multi-junction solar cells, the direct terrestrial sunlight spectrum (AM1.5D), defined for a zenith angle of 48.2° representing the average conditions of the United States, is split between each sub-cell in this triple junction design as shown in Figure13. The bandwidth of absorption and internal quantum efficiency extends in both the UV and IR directions, ranging from 300~1800 nm. Secondly, for the concentrator multi-junction solar cells, light is incident upon the cell over a wide angular range, introducing an additional dimension for optimization. Thirdly, solar cells are required to operate for 20~30 years. Materials must not be modified or damaged by long-term exposure to UV light or large periodic changes in temperature and humidity. Furthermore, variations in the temperature-dependence of the refractive index of each layer will lead to a temperature-dependent transmission spectrum which may affect the performance of multi-junction solar cells. Finally, these ARC layers should be deposited inexpensively over large areas, together in a single coating chamber, and at low temperatures to minimize impact on the solar cell performance.

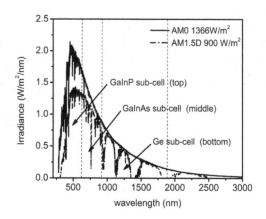

Figure 13. The wavelength versus irradiance spectrum.

In the triple junction solar cells, the window layer of the top cell, AlInP, was considered for the ARC designs, using a structure (air, ARC layer(s), AlInP) with direct normal incidence AM1.5D sunlight. Figure14 shows the reflection spectra for two-layer material combinations commonly used for antireflection coatings. Both Al_2O_3/TiO_2 and SiO_2/TiO_2 offer coating solutions using practical deposition equipment. Commercially deposited Al_2O_3/TiO_2 coatings have shown a 30-35% improvement in the Isc and a corresponding smaller increase to Voc when compared with uncoated devices.

Figs.15 shows the improvement in the external quantum efficiency (EQE) of the cells with Al_2O_3/TiO_2 coatings. It can be seen that an improvement in the EQE (AM1.5D) of the top and middle cells is from 65% to 88%. Figure16 shows the improvement in the optical and electrical properties of the samples above. It can be seen that the Isc under 1000 suns is improved from 10.27 to 13.79 A, and the improvement in the Isc is 34.3%; the Voc also has a small increase of 0.03V. The FF shows an obvious decrease of 1.9% because of more ohmic loss with higher Isc. The efficiency of the samples with Al_2O_3/TiO_2 coatings combination increases from 29.33% to 39.30%, a 34.0% improvement.

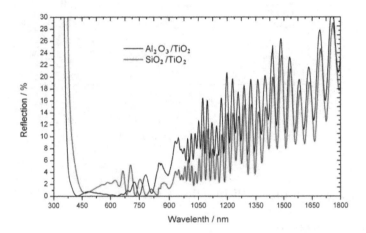

Figure 14. The reflection spectra of the solar cells with ARC.

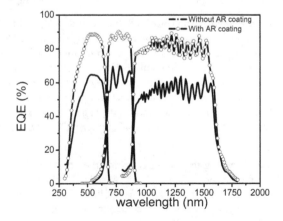

Figure 15. The EQE of the cells with and without ARC.

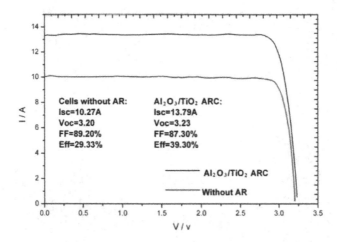

Figure 16. The optical and electrical properties of the cells with and without ARC.

4. The reliability of the multi-junction solar cell

4.1. Accelerated aging tests

Under high concentration of several hundreds or even thousands suns, multi-junction solar cells will suffer high temperature and high current density, which are challenging the reliability of these devices [16]. To obtain the approval from CPV customers, it is necessary to demonstrate the reliability of multi-junction solar cells operating under high concentration.

A new certification standard, namely the IEC62108, has been developed in which the procedure for qualifying CPV systems and assemblies is described. The IEC62108 is currently the only international standard on assessment of high concentration solar receivers and modules, which specifies the minimum requirements for the design qualification and the type approval of concentrator solar cells, and which gives the corresponding test procedures for each test sample, such as outdoor exposure test, electrical performance measurement, electrical test, irradiation test, and mechanical load test. After passing the IEC62108 certification, both the modules and assemblies can be suitable for long-term (25 years) operation in general open-air climates.

The purpose of the thermal cycling test is to determine the ability of the receivers to withstand thermal mismatch, fatigue, and other stresses caused by rapid, non-uniform or repeated changes of temperature. This test is vital to the reliability of concentrator solar cells, since generally these devices have to operate at high concentration of more than 1000 suns, high operation current density of more than 10A/cm^2, high operation temperature of more than 60 °C and large temperature difference between day and night.

In order to simulate the real operating conditions, IEC 62108 requires that during the process of thermal cycling test for concentrator solar cells carried out in the oven, a current should be flowing through the chips. Table 6 shows the three optional conditions. In principle, the temperature and the current injection time of cells are required to be accurately monitored during thermal cycling test. However, it is very difficult to monitor the real temperature of cells in real operating conditions, because a high electric current passing through the cells can lead to differences in temperature among the cells, the heat sink and the oven.

Option	Maximum cell temperature	Total cycles	Applied current
TCA-1	85 °C	1000	Apply 1.25xIsc when T "/ 25°C, cycle speed is 10 electrical/thermal
TCA-2	110 °C	500	Apply 1.25xIsc when T "/ 25°C, cycle speed is 10 electrical/thermal
TCA-3	65 °C	2000	Apply 1.25xIsc when T "/ 25°C, cycle speed is 10 electrical/thermal

Table 6. The options of thermal cycling test from IEC 62108.

Using the thermal cycling test condition of TCA-1 from table 6, the cell temperature is controlled between -40 °C and 85 °C. A dwell time of 10 min of the high and low temperatures is required. The cycling period and frequency are 120 minutes and 12 cycles per day, respectively. In one thermal cycle, a specific current level of 7A is periodically on and off for 10 cycles, when the cell temperature is above 25 °C. In order to illustrate the changes of electrical performance of test samples, control samples are chosen and measured under the similar test condition. By this method, test condition variables are self-corrected, and the complex translation procedures are eliminated. Finally, the relative power Pr and relative power degradation Prd are defined as follows:

$$P_r = \frac{P_m}{P_{mc}} \times 100\%$$ (25)

$$P_{rd} = \frac{P_{ri} - P_{rf}}{P_{ri}} \times 100\%$$ (26)

where Pm is the test sample's maximum power, Pmc is the control sample's maximum power measured at the similar condition as Pm, and Prf and Pri are the relative powers measured after and before the given test, respectively.

For comparison, eight San'an company's cells and eight B-company's cells were tested together. Tables 7 and 8 show the relative power degradation of San'an Company's and B-company's receiver samples after different numbers of thermal cycles, respectively. The output powers gradually decrease with the increasing thermal cycles due to the samples' degradation. It is found that the relative power degradations of tested samples are within 10%. The degradation is believed to be responsible for the perimeter degradation [16, 17]. According to González et al., the arbitrary definition of device failure is a 10% of power loss, so the majority of test samples do not have failure, except the c B-company's receiver #56, the relative power degradation of which is from 12.85% after 560 thermal cycles to 14.89% after 1000 thermal cycles. Besides, from visual inspection on these samples, the DBCs soldered on alumina substrates are not peeled off after the 1000 thermal cycles, which indicates that it is suitable for long-term (~25 years) operation in general open-air climates.

Serial sample	0 cycle	360 cycles	560 cycles	760 cycles	1000 cycles
#182B5	0.00%	-2.47%	-4.00%	-5.56%	-8.02%
#183D1	0.00%	-0.14%	-3.19%	-3.88%	-5.44%
#182D5	0.00%	-3.66%	-4.78%	-5.82%	-8.21%
#183B1	0.00%	-2.48%	-4.83%	-6.33%	-8.25%
#182D1	0.00%	-4.62%	-5.38%	-6.21%	-7.95%
#183B6	0.00%	-5.85%	-6.08%	-5.84%	-7.09%
#183A4	0.00%	-5.02%	-5.70%	-6.97%	-7.96%
#183D5	0.00%	-5.47%	-5.83%	-6.06%	-7.37%

Table 7. Relative power degradation of San'an Company's receiver samples after different numbers of thermal cycles.

Serial sample	0 cycle	360 cycles	560 cycles	760 cycles	1000 cycles
#112	0.00%	-5.93%	-6.83%	-6.84%	-7.72%
#44	0.00%	-4.10%	-5.60%	-6.12%	-8.11%
#56	0.00%	-8.02%	-12.85%	-13.05%	-14.89%
#94	0.00%	-2.76%	-4.94%	-5.19%	-6.78%
#78	0.00%	-5.17%	-6.39%	-6.76%	-7.36%
#90	0.00%	-7.14%	-7.75%	-8.17%	-8.32%
#97	0.00%	-3.19%	-3.38%	-4.38%	-7.42%
#136	0.00%	-6.30%	-6.69%	-7.05%	-9.34%

Table 8. Relative power degradation of B-company's receiver samples with different numbers of thermal cycles.

In conclusion, high concentration multi-junction solar cells are still at an early stage of technological development, and thus it is necessary to demonstrate the reliability of these solar cells before their industrialization. Accelerated aging test is a necessary tool to demonstrate the reliability of concentration photovoltaic solar cells, which is expected to be working for no less than 25 years. According to the requirements from IEC 62108, this paper presents the reliability results from thermal cycling tests performed on San'an company's high concentration solar cells. We find that the light emitting intensity and the relative power degradation of San'an company's receivers are similar to that of B-company's receivers.

4.2. Discussion on outdoor power plant performance

Concentrated photovoltaic (CPV) system is usually located in sunny places for large-scale photovoltaic (PV) power station with installation capacity of 1~1000 megawatt (MW). It is composed of Fresnel lenses to concentrate, III-V multi-junction solar cells, polar axis type or pedestal type tracking system and integrated control method. By focusing sunlight onto high-efficiency solar cells, CPV is able to use fewer solar cells than traditional photovoltaic power. Since CPV has a high power-generating capacity with movable parts, easy to manufacture and to maintain, it is very suitable for a large scale PV power station.

According to the CPV Consortium, "CPV, with its higher efficiency delivers higher energy production per megawatt installed, provides the lowest cost of solar energy in high solar regions of the world. The technology is in its early stage with significant headroom for future innovation, and it has the ability to ramp to gigawatts of production very rapidly. Many of the limitations for PV in the past are overcome by advances in CPV technology."As of 2011, the global bases of installed CPV produced totally just 60 megawatts, according to the CPV Consortium. The organization predicts that capacity will rise to 275 megawatts by the end of 2012, 650 megawatts by the end of 2013, 1,100 megawatts by end of 2014 and 1,500 megawatts by the end of 2015.

World-widely, 40 MW Amonix power plants will be installed from 2012 on, at the same time 32.7 MW power plant located at Alamosa Colorado was measured during the week of March 2012. ISFOC (Institute of Concentration Photovoltaic Systems) main goal is to promote the CPV industrialization. For this purpose, ISFOC has made the installation of CPV Plants, up to 2.7 MW, all over the region of Castilla la Mancha. A lot of CPV power plants will be installed in near future without being introduced more. However, focusing on China, the relative long history of advanced CPV technology development, the years' experience of power plant operation, mature systems with high performance and reliability, the leading position of the western participants will set up a benchmark in the field and gain more attention and more shares from Chinese CPV market. For a few domestic CPV companies with installation records, further efforts are required to improve the performance and reliability of CPV products, to lower the cost by setting up complete supply chains in CPV industry, to facilitate the utilization of abundant solar resources from the north and west to the south and east via setting up transmission networks, so that a Chinese CPV market can be actually initiated, developed and matured.

The largest CPV power plant project in China was assembled at Golmud, Qinghai province by Suncore Photovoltaic Technology Co., Ltd, with the capacity of more than 50 MW. Suncore is a Sino-US joint venture established by San'an and Emcore. 1 MW of the project using 500 suns terrestrial system and 2MW using 1000 suns terrestrial system has been finished, as shown in Figure 17. Conversion efficiency of 500 suns and 1000 suns terrestrial system can reach as high as 25% and 28.5%, respectively.

Figure 17. power plant installed at Golmud Qinghai province China.

The direct normal insolation (DNI) distribution of the local environment and the mapping of China were displayed Figure 18 (a) and (b). The I-V curves of 227 receivers using 500 suns terrestrial system module tested outdoor was shown in Figure 19. One can see that the efficiency could reach as high as 24.03% at the condition of much dust on the surface of the Fresnel lens, which affecting the light transmittance. Therefore, the actual efficiency should be high than this nominal value.

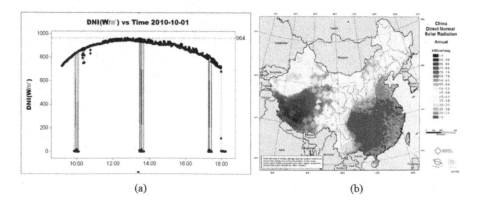

(a)   (b)

Figure 18. The DNI distribution of the whole day in Golmud (a), and the annual average direct normal insolation (DNI) GIS data at 40km resolution for China (b) from NREL.

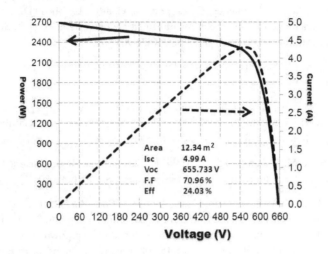

Figure 19. I-V curves of the 227 Receivers module tested outdoor.

Acknowledgements

This work was supported by a foundation from the National High Technology Research and Development Program (863 program) of China. (No. 2012AA051402).

Author details

Gui jiang Lin[1], Jingfeng Bi[1], Minghui Song[1], Jianqing Liu[1], Weiping Xiong[1] and Meichun Huang[2*]

*Address all correspondence to: mchuang@xmu.edu.cn

1 Xiamen San'an Optoelectronics Co., Ltd., China

2 Department of Physics, Xiamen University, China

References

[1] Green, M. A. (2005). *Third generation photovoltics*, Berlin, Springer.

[2] Luque, A., Martı´, A., Stanley, C., Lo´pez, N., Cuadra, L., Zhou, D., & Mc Kee, A. (2004). General equivalent circuit for intermediate band devices: Potentials, currents and electroluminescence. *J. Appl. Phys.*, 96(03), 903-909.

[3] King, R. R. (2008). Multi-junction solar cells: Record breakers. *Nature Photonics*, 2, 284-286.

[4] King, R. R., Law, D. C., Edmondson, K. M., Fetzer, C. M., Kinsey, G. S., Yoon, H., Sherif, R. A., & Karam, N. H. (2007). 40% efficient metamorphic GaInP/GaInAs/Ge multi-junction solar cells. *Appl. Phys. Lett.*, 90(18), 3516-3518.

[5] Yamaguchi, M., Takamoto, T., & Araki, K. (2006). Super high-efficiency multi-junction and concentrator solar cells. *Sol. Energy Mater. Sol. Cells*, 90-3068.

[6] Daniel, C. L., King, R. R., Yoon, H. M., Archer, J., Boca, A., Fetzer, C. M., Mesropian, S., Isshiki, T., Haddad, M., Edmondson, K. M., Bhusari, D., Yen, J., Sherif, R. A., Atwater, H. A., & Karama, N. H. (2010). Future technology pathways of terrestrial III-V multi-junction solar cells for concentrator photovoltaic systems. *Solar Energy Materials & Solar Cells*, 94, 1314-1318.

[7] Geisz, J. F., Kurtz, S., Wanlass, M. W., Ward, J. S., Duda, A., Friedman, D. J., Olson, J. M., Mc Mahon, W. E., Moriarty, T. E., & Kiehl, J. T. (2007). High-efficiency GaInP/ GaAs/InGaAs triple-junction solar cells grown inverted with a metamorphic bottom junction. *Appl. Phys. Lett.*, 91(02), 3502-3504.

[8] Guter, W., Schöne, J., Philipps, S. P., Steiner, M., Siefer, G., Wekkeli, A., Welser, E., Oliva, E., Bett, A. W., & Dimroth, F. (2009). Current-matched triple-junction solar cell reaching 41.1% conversion efficiency under concentrated sunlight. *Appl. Phys. Lett.*, 94(22), 3504-3506.

[9] Fahrenbruch, A. L., & Bube, R. H. (1983). *Fundamentals of Solar Cells Photovoltaic Solar Energy Conversion*, New York, Academic Press.

[10] Würfel, P. (2005). *Physics of solar cells: From principles to new concepts. Verlag GmbH & Co KGaA*, Weinheim, Wiley-VCH.

[11] Olson, J. M., Ahrenkiel, R. K., Dunlavy, D. J., Keyes, B., & Kibbler, A. E. (1989). Ultralow recombination velocity at Ga0.5In0.5P/GaAs heterointerfaces. *Appl. Phys. Lett.*, 55, 1208-1210.

[12] Palik, E. D. (1991). *Handbook of Optical Constants of Soilds II*, San Diego, USA, Academic Press.

[13] Kalyuzhnyy, N. A., Gudovskikh, A. S., Evstropov, V. V., Lantratov, V. M., Mintairov, S. A., Timoshina, K. H., Shvarts, M. Z., & Andreev, V. M. (2010). Germanium Subcells for Multi-junction GaInP/GaInAs/Ge Solar Cells. *Semiconductors*, 44(11), 1520-1528.

[14] Ebert, C., Pulwin, Z., Byrnes, D., Paranjpe, A., & Zhang, W. (2010). Tellurium doping of InGaP for tunnel junction applications in triple junction solar cells. *Journal of Crystal Growth, 315,* 61-63.

[15] Garcia, I., Rey-Stolle, I., Algora, C., Stolz, W., & Volz, K. (2008). Influence of GaInP ordering on the electronic quality of concentrator solar cells. *Journal of Crystal Growth,* 310, 5209-5213.

[16] González, J. R., Vázquez, M., Núñez, N., Algora, C., Rey-Stolle, I., & Galiana, B. (2009). Reliability analysis of temperature step-stress tests on III-V high concentration solar cells. *MICROELECTRONICS RELIABILIBTY,* 49, 673-680.

[17] Algora, C. (2010). Reliability of III-V concentrator solar cells. *MICROELECTRONICS RELIABILIBTY,* 50, 1193-1198.

Chapter 8

Use of Optoelectronic Plethysmography in Pulmonary Rehabilitation and Thoracic Surgery

Giulia Innocenti Bruni, Francesco Gigliotti and Giorgio Scano

Additional information is available at the end of the chapter

1. Introduction

1.1. Methods

OEP system is an optoelectronic device able to track the three-dimensional co-ordinates of a number of reflecting markers placed non-invasively on the skin of the subject [1-4]. A variable number of markers (89 in the model used for respiratory acquisition in seated position) is placed on the thoraco-abdominal surface; each marker is a half plastic sphere coated with a reflective paper. Two TV Sensors 2008, cameras are needed to reconstruct the X-Y-Z co-ordinates of each marker, so for the seated position six cameras are required. Each camera is equipped with an infra-red ring flash. This source of illumination, which is not visible, is not disturbing and lets the system also operate in the dark. The infra-red beam, emitted by the flashes, is reflected by each marker and acquired by the cameras with a maximal sampling rate of 100 Hz. The signal is then processed by a PC board able to combine the signal coming from the cameras and to return, frame by frame, the three-dimensional co-ordinates of each marker. The process is simultaneously carried out for the six TV cameras needed for the seated respiratory model. Acquired data need a further operation called 'tracking' that is necessary to exclude possible phantom reflections and/or to reconstruct possible lost markers (this could happens sometimes during very fast manoeuvres such exercise); at this time the obtained files contain the X-Y-Z co-ordinates of each marker during the recorded manoeuvre, then data are stored on the PC hard disk. The spatial accuracy for each marker's position is about 0.2 mm [1]. Volumes for each compartment is calculated by constructing a triangulation over the surface obtained volume from the X-Y-Z co-ordinates of the markers and then using Gauss's theorem to convert the volume integral to an integral over this surface [2]. The number and the position of used markers depends on the thoraco-abdominal model chosen. As proposed by Ward & Macklem [5] we use a three compartment chest wall model: the upper rib cage, lower rib cage and abdomen. Due to the fact that the upper portion of the rib cage is exposed to pleural pressure whereas the lower portion is affected by abdominal pressure, a model able to dynamically return changes in volume of each compartment and, as a sum, of the entire chest wall has been developed [2]. The number of used markers is 89, 42 placed on the front and 47 on the back of the subject.

To measure the volume of chest wall compartments from surface markers we define: 1) the boundaries of the upper rib cage as extending from the clavicles to a line extending transversely around the thorax at the level of the xiphoid process (corresponding to the top of the area of the apposition of the diaphragm to the rib cage at end expiratory lung volume in sitting posture, confirmed by percussion); 2) the boundaries of lower rib cage as extending from this line to the costal margin anteriorly down from the xiphosternum, and to the level of the lowest point of the lower costal margin posteriorly; and 3) the boundaries of abdomen as extending caudally from the lower rib cage to the level of the anterior superior iliac crest. The markers are placed circumferentially in seven horizontal rows between the clavicles and the anterior superior iliac spine. Along the horizontal rows the markers are arranged anteriorly and posteriorly in five vertical rows, and there is an additional bilateral row in the midaxillary line. The anatomical landmarks for the horizontal rows are: 1) the clavicular line; 2) the manubrio-sternal joint; 3) the nipples (~ 5 ribs); 4) the xiphoid process; 5) the lower costal margin (10th rib in the midaxillary line); 6) umbilicus; 7) anterior superior iliac spine. The landmarks for the vertical rows are: 1) the midlines; 2) both anterior and posterior axillary lines; 3) the midpoint of the interval between the midline and the anterior axillary line, and the midpoint of the interval between the midline and the posterior axillary line; 4) the midaxillary lines. An extra marker is added bilaterally at the midpoint between the xiphoid and the most lateral portion of the 10th rib to provide better detail of the costal margin; two markers are added in the region overlying the lung-apposed rib cage and in the corresponding posterior position. This marker configuration has previously been validated in normal subjects, along with a sensitivity analysis which assesses accuracy in estimating change in lung volume as a function of marker number and position [2]. When compared with the gold standard (water sealed spirometer) the accuracy in the volume change measurements of the 89 markers model is very high, showing volume differences smaller than 5% [2].

2. Exercise limitation and breathlessness in patients with Chronic Obstructive Pulmonary Disease (COPD)

Dynamic hyperinflation (DH) is supposed to be the most important factor limiting exercise and contributing to dyspnea by restrictive constraints to volume expansion in patients with COPD [6]. Indirect evidence of the importance of DH has been provided by studies that have demonstrated that pharmacological treatment [7,8], and lung volume reduction surgery [9] explain in part the improvement in exercise performance and dyspnea by reducing DH in these patients. It has recently been found, however, that different patterns of chest wall kinematics may or may not be associated with different exercise performance in COPD patients [10,11]. There is little data available indicating that these patients may dynamically hyperinflate or deflate chest wall compartments during cycling while breathing air [10,12] or with oxygen supplementation [11]. As yet the contribution of reducing lung volume to dyspnea relief remains uncertain [11,13-15] in exercising COPD patients. It also remains to be determined whether changes in operational chest wall volumes substantially affect the response to endurance exercise rehabilitation programs. It should be remembered that (i) an increase in end-expiratory-volume of the chest wall constrains the potential for the tidal volume to increase; thus exacerbating the sensation of dyspnea; (ii) on the other hand, shifting abdominal volumes towards a lower operational point might not be able to reduce restrictive constraints on volume displacement if the rib cage dynamically hyperinflates. Arguably, rib cage hyperinflation would result in a higher volumetric load to the intercostal inspiratory muscles [16] and a higher sensory perception of dyspnea [17]; (iii) the possibility that abdominal deflation contributes *per se* to dyspnea should not be disregarded [16]. Evidence has indeed been provided that a decrease in abdominal volume resulting from increased abdominal muscle activity as soon as exercise starts even at minimal work rate [4] may contribute *per se* to increasing the work of breathing [10] and breathlessness [16], to reducing venous return and cardiac output [18], and to decreasing exercise capacity [4] in patients with COPD.

Dyspneic patients with COPD who are markedly hyperinflated are considered especially likely to display abnormalities in rib cage motion such as a paradoxical (inward) inspiratory movement of their lower rib cage [19-22]. Studies in healthy humans have led to the hypothesis that the primary mechanism of abnormal chest wall motion in patients with COPD is probably an abnormal alteration of forces applied to chest wall compartments [3,23] and an increase in airway resistance [24]. Chihara *et al.* [23] have speculated that when rib cage distortion is present, greater degree of recruitment of inspiratory rib cage muscles and greater predisposition to dyspnea for a given load and strength do occur. On the other hand, the role of hyperinflation on abnormal chest movement is questionable in healthy subjects [24]. Accordingly, it has recently been shown that paradoxical movement of the lower rib cage cannot be fully explained by static lung hyperinflation [19] or dynamic rib cage hyperinflation [25] in patients with COPD. By contrast, Aliverti *et al.* [26] have shown that lower rib cage paradox results in an early onset of dynamic hyperinflation as a likely explanation for the increased exertional breathlessness in these patients. Nonetheless, the link between changes in operational lung volumes and exertional breathlessness has not been definitely established in normoxic COPD patients [13,14,27].

Now the questions arise: does exercise reconditioning reduce rib cage distortion, and, if any, does rib cage distortion contribute to restoring exercise capacity and to relieving breathlessness Does exercise reconditioning relieve dyspnea regardless of whether compartmental chest wall volumes are shifted toward upper or lower operational points?

3. How Optoelectronic Plethysmography (OEP) can help answer the above questions

The use of Optoelectronic Plethysmography (OEP) has allowed us to understand some of the mechanisms underlying the efficacy of rehabilitative treatment in patients with COPD. Rehabilitation interventions such as oxygen supplementation reduce ventilation and the rate of dynamic hyperinflation, but whether and to what extent reduction in lung volume contributes to dyspnea relief remains uncertain in these patients [13,14,27]. Innocenti Bruni et al. [11] tried to (i) determine whether and how hyperoxia would affect exercise dyspnea, chest wall dynamic hyperinflation, and rib cage distortion in normoxic COPD patients, and (ii) investigate whether these phenomena are interrelated. It was speculated that they are not, based on the following observations: (i) significant dyspnea relief and improvement in exercise endurance can occur even in the absence of an effect on dynamic lung hyperinflation [27]; (ii) externally imposed expiratory flow limitation is associated with no rib cage distortion during strenuous incremental exercise, with indexes of hyperinflation not being correlated with dyspnea [16]; (iii) end-expiratory-chest wall-volume may either increase or decrease during exercise in patients with COPD, with those who hyperinflate being as breathless as those who do not [10]; (iv) a similar level of dyspnea is associated with different increases in chest wall dynamic hyperinflation at the limits of exercise tolerance [28]. The volume of chest wall (Vcw) and its compartments: the upper rib cage (Vrcp), lower rib cage (Vrca), and abdomen (Vab) were evaluated by OEP in 16 patients breathing either room air or 50% supplemental O_2 at 75% of peak exercise in randomized order; rib cage distortion was assessed by measuring the phase angle shift between Vrcp and Vrca. Ten patients increased end-expiratory Vcw (Vcw,ee) on air. In 7 *hyperinflators* and 3 *non-hyperinflators* the lower rib cage paradoxed inward during inspiration with a phase angle of 63.4° (30.7) compared with a normal phase angle of 16.1° (2.3) recorded in patients without rib cage distortion. Dyspnea by a modified Borg scale from zero (no dyspnea), to ten (maximum dyspnea) averaged 8.2 and 9 at end-exercise on air in patients with and without rib cage distortion, respectively. At iso-time during exercise with oxygen, dyspnea relief was associated with a decrease in ventilation regardless of whether patients distorted the rib cage, dynamically hyperinflated or deflated the chest wall. (Fig 1).

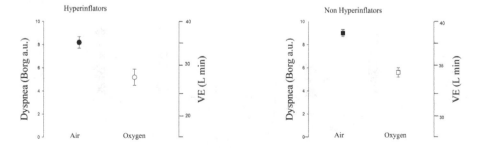

Figure 1. Oxygen supplementation decreses ventilation and dyspnea at isotime during constant load cycling exercise.

OEP allows us to demonstrate that dyspnea, chest wall dynamic hyperinflation, and rib cage distortion are not interrelated phenomena.

Georgiadou *et al.* [12] studied the effect of pulmonary rehabilitation on the regulation of total chest wall and compartmental volumes during exercise in patients with COPD. Twenty patients undertook high-intensity exercise 3 days week[-1] for 12 weeks. Before and after rehabilitation, the changes in chest wall (cw) volumes at the end of expiration (Vcw,ee) and inspiration (Vcw,ei) were computed by OEP during incremental exercise to the limit of tolerance (W_{peak}). Rehabilitation significantly improved W_{peak} In the post-rehabilitation period and at identical work rates, significant reductions were observed in minute ventilation, breathing frequency and Vcw,ee and Vcw,ei. Inspiratory reserve volume was significantly increased. Volume reductions were attributed to significant changes in abdominal Vcw,ee and Vcw,ei. The improvement in W_{peak} was similar in patients who progressively hyperinflated during exercise and those who did not. The authors concluded that pulmonary rehabilitation lowers chest wall volumes during exercise by decreasing the abdominal volumes.

The study indicates that improvement in exercise capacity following rehabilitation is independent of the pattern of exercise-induced dynamic hyperinflation.

Preliminary laboratory data indicate that OEP substantially assists in clarifying the link between chest wall dynamic hyperinflation and breathlessness following pulmonary rehabilitation. The volume of the chest wall and its compartments were evaluated in 14 patients by OEP during constant load cycle exercise before and after pulmonary rehabilitation. Prior to rehabilitation exercise increased end-expiratory chest wall volumes in eight patients, but deflated the chest wall in six [11]. Rehabilitation increased exercise endurance. Relief in both dyspnea and leg effort at iso-time were associated with a decrease in ventilation regardless of whether patients hyperinflated or not. Also, the effect of pulmonary rehabilitation on rib cage distortion and dyspnea were independent of each other (Fig 2).

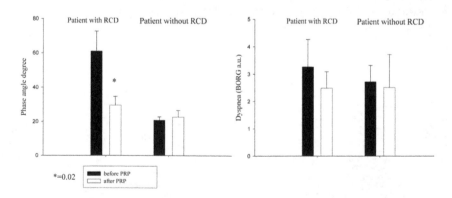

Figure 2. Effect of pulmonary rehabilitation (PRP) on phase angle and dyspnea in patients with and without ribcage distortion (RCD).

These data suggest that pulmonary rehabilitation reduces dyspnea regardless of rib cage distortion and dynamic chest wall hyperinflation.

Many COPD patients complain of severe dyspnea while performing simple daily-life activities using their arms. The increased demand during simple arm elevation may play a role in the development of dyspnea and in the limitation that is frequently reported by these patients when performing activities involving their arms [29,30]. Unsupported arm exercise training (UAET) is increasingly recognized as an important component of pulmonary rehabilitation in these patients [31]. Although some studies have demonstrated improvement in unsupported arm exercise after UAET [32-34], suggesting that the test can be sensitive to changes in arm exercise capacity, the impact of upper extremity training on arm exercise related-dyspnea and fatigue remains unclear [35-38] or undemonstrated [32,38-40]. Surprisingly, few studies [32,35,37-39] have investigated the effect of upper extremity training on ratings of perceived dyspnea by applying psychophysical methods, that is, the quantitative study of the relationship between stimuli and evoked conscious sensory responses. On this basis we have recently demonstrated that neither chest wall dynamic hyperinflation nor dyssynchronous breathing *per se* are the major contributors to dyspnea during unsupported arm exercise in COPD patients [25]. Using the same approach we have recently tried to document the impact of arm training on arm exercise-related perceptions. The finding that before rehabilitation patients stop arm exercise namely because of arm symptoms, makes a case for the excessive effort felt by subjects being elicited by arm/torso afferent information (from the muscles performing the excessive effort) conveyed to the motor-sensory cortex [25].

These findings may explain why even a very small decrease in ventilatory demand, reflective of a decrease in central motor output to ribcage/torso muscles, has a salutary effect on arm symptoms during arm training in patients with COPD [41].

OEP has also helped to clarify mechanisms by which some techniques of pulmonary rehabilitation such as breathing retraining, namely "pursed lip breathing" (PLB), act in reducing the sensation of dyspnea. Bianchi *et al.* [42] hypothesized that the effect of PLB on breathlessness relies on its deflationary effects on the chest wall. They found that patients exhibited a significant reduction in end-expiratory volume of the chest wall (Vcw,ee) and a significant increase in end-inspiratory volume of the chest wall in comparison with spontaneous breathing. In a stepwise multiple regression analysis, a decrease in end expiratory volume of the chest wall accounted for 27% of the variability in the Borg score.

These data indicate that by lengthening the expiratory time, PLB deflates the chest wall and reduces dyspnea.

In a further paper Bianchi *et al.* [43] identified the reasons why some patients benefit from PLB while others do not. The OEP analysis of chest wall kinematics shows why not all patients with COPD obtain symptom relief from PLB at rest. The most severely affected patients who deflate the chest wall during volitional PLB reported improvement in their sensation of breathlessness. This was not the case in the group who hyperinflated during PLB.

4. Comparing OEP with spirometric operational volumes

OEP may provide complementary information on operational volumes to that provided by spirometry. Vogiatzis *et al.* [28] found a good relationship between changes in inspiratory capacity (ΔICpn) and changes in end expiratory chest wall volume (ΔVcw,ee). By contrast we have not found any significant relationship between the two measurements (Fig 3).

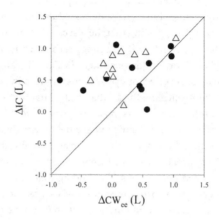

Figure 3. Plots of change in inspiratory capacity (IC) vs change in end-expiratory-chest wall-volumes (CW$_{ee}$) from rest to end exercise, before (closed circles) and after (triangles) pulmonary rehabilitation. Continuous line is the identity line.

The decrease in ICpn is much greater than the increase in Vcw,ee in most patients. The reasons for this discrepancy are probably due to: i) error measurements with the pneumotachograph possibly linked to leakage and elevation of temperature in the system, and to spirometric drift resulting in spurious increments or decrements in volume measurements; ii) spirometry measures the volume of the gas entering or leaving the lungs at the mouth, while OEP measures the volume of the trunk which includes changes in gas volume, gas compression and blood volume shifts [16]. Arguably, activity of the abdominal muscles producing various amounts of gas compression and blood shifts might account for the prevalence of one method over the other. For instance, high gas compression and blood shift would result in a greater decrease in Vcw,ee than an increase in the next ICpn manoeuvre [44]. It has been postulated that OEP would not detect 89% of the reduction in inspiratory capacity measured with spirometry in some conditions [44].

5. OEP and reparative deformity of the rib cage

Pectus excavatum, the most common congenital chest wall deformity, is characterized by a depression of the anterior chest wall and sternum. Some patients will develop cardiopulmonary symptoms for the first time as adolescents while others will experience a worsening of the symptoms they have endured for years. A minimally invasive technique for repair described by Nuss *et al.* [45,46] involves the placement of substernal concave bar(s) that will be rotated to elevate the sternum outward. The bar is left in place for 2-3 years while the anterior chest wall remodels. The chest wall is primarily involved when there are respiratory abnormalities, so the effect of repair should be assessed mainly by observing chest wall kinematics and possibly chest wall mechanics in pectus excavatum patients. A previous study carried out in adolescents with mild restrictive defect has shown that abnormalities in chest wall kinematics during maximal voluntary ventilation are not correlated with the computed tomography scan severity index, indicating the contribution of chest wall kinematics to clinical evaluation of pectus excavatum patients [47]. Should we wait 2-3 years before assessing repair effects (if any) on chest wall kinematics? Can the Nuss procedure influence timing, and kinematics of the chest wall and rib cage configuration in otherwise healthy subjects? Binazzi *et al.* [48] postulated that the repair effect based on increased chest wall end expiratory volume does not affect chest wall displacement and dynamic configuration of the rib cage. By using OEP they provided a quantitative description of chest wall kinematics before and 6 months after the Nuss procedure at rest and during maximal voluntary ventilation in 13 subjects with pectus excavatum. An average 11% increase in chest wall volume was accommodated within the upper rib cage and to a lesser extent within the abdomen and lower rib cage. Tidal volumes did not significantly change during the study period. The repair effect on chest wall kinematics did not correlate with the Haller index of deformity at baseline.

These data indicate that six months of the Nuss procedure do increase chest wall volume without affecting chest wall displacement and rib cage configuration.

6. Conclusion

In conclusion, we and others have shown that use of OEP can demonstrate the following: (i) dynamic hyperinflation of the chest wall may not necessarily be the principal reason for exercising limitation and breathlessness in COPD patients; (ii) pulmonary rehabilitation improves COPD patients' endurance and exercise-related perceptions regardless of changes in chest wall kinematics; (iii) in contrast with what is commonly believed, chest wall dynamic hyperinflation may have a salutary mechanical effect in patients with expiratory flow limitation and dynamic hyperinflation, who increase functional residual capacity because of achieving more tidal expiratory flow; (iv) OEP provides complementary information on operationl volumes to that provided by spirometry.

Finally, there are very few reports on the use of OEP in pulmonary rehabilitation and thoracic surgery in patients with chronic respiratory disease other than COPD. We hope that the results presented here will stimulate new contributions on this topic.

Author details

Giulia Innocenti Bruni[1], Francesco Gigliotti[1] and Giorgio Scano[1,2]

1 Section of Respiratory Rehabilitation, Fondazione Don C. Gnocchi, Florence, Italy

2 Department of Internal Medicine, Section of Immunology and Respiratory Medicine, University of Florence, Florence, Italy

References

[1] Pedotti A, Ferrigno G. Opto-electronics based systems. In Three-Dimensional Analysis of Human Movement, Human Kinetics, 1st Ed.; Allard P, Stokes IA, Bianchi JP. Eds.; Human Kinetics Publishers: Champaign, USA, 1995; pp. 57-78.

[2] Cala SJ, Kenyon C.M, Ferrigno G, Carnevali P, Aliverti A, Pedotti A,Macklem PTRochester DF. Chest wall and lung volume estimation by optical reflectance motion analysis. Journal of applied Phisiology. 1996; 81:2680-2689.

[3] Kenyon CM, Cala SJ, Yan S, et al. Rib cage mechanics during quiet breathing and exercise in humans. Journal of Applied Physiology. 1997;83:1242-55.

[4] Aliverti A, Cala SJ, Duranti R, Ferrigno G, Kenyon CM, Pedotti A, Scano G, Sliwinski P, Macklem PT, Yan S. Human respiratory muscle action and control during exercise. Journal of Applied Physiology 1997;83:1256-1269

[5] Ward ME,Ward JW, Macklem PT. Analysis of human chest wall motion using a two compartment rib-cage model.Journal of Applied Physiology . 1992; 72:1338-1347.

[6] O' Donnell DE, Revill SM, Webb KA. Dynamic hyperinflation and exercise intolerance in COPD. American Journal of Respiratory and Critic Care Medicine 2001;164:770-777.

[7] O'Donnell DE, Voduc N, Fitzpatrick M, Webb KA. Effect of salmeterol on the ventilatory response to exercise in chronic obstructive disease. European Respiratory Journal 2004;24:86-94.

[8] Belman MJ, Botnick WC, Shin JW. Inhaled bronchodilators reduce dynamic hyperinflation during exercise in patients with chronic obstructive pulmonary disease. American Journal of Respiratory and Critic Care Medicine 1996;153:967-975.

[9] Martinez FJ, de Oca MM, Whyte RI, Stetz J, Gay SE, Celli BR. Lung-volume reduction improves dyspnea, dynamic hyperinflation, and respiratory muscle function. American Journal of Respiratory and Critic Care Medicine 1997;155(6):1984-90.

[10] Aliverti A, Stevenson N, Dellacà RL, Lo Mauro A, Pedotti A, Calverley PMA. Regional chest wall volumes during exercise in chronic obstructive pulmonary disease. Thorax 2004;59:210-216.

[11] Innocenti Bruni G, Gigliotti F, Binazzi B, Romagnoli I, Duranti R, Scano G. Dyspnea, chest wall hyperinflation, rib cage distortion in exercising COPD patients. Medicine and Science in Sport & Exercise 2012. DOI:10.1249

[12] Georgiadou O, Vogiatzis I, Stratakos G, Koutsoukou A, Golemati S, Aliverti A, Roussos C, Zakynthinos S. Effects of rehabilitation on chest wall volume regulation during exercise in COPD patients. European Respiratory Journal 2007; 29:284-291.

[13] O'Donnell DE, Bain DJ, Webb K. Factors contributing to relief of exertional breathlessness during hyperoxia in chronic air flow limitation. American Journal of Respiratory and Critic Care Medicine 1997;155:530-535.

[14] Somfay A, Porszasz J, Lee SM, Casaburi R. Dose-response effect of oxygen on hyperinflation and exercise endurance in non hypoxaemic COPD patients. European Respiratory Journal 2001;18:77-84.

[15] Gigliotti F, Coli C, Bianchi R, Romagnoli I, Lanini B, Binazzi B, Scano G. Exercise training improves exertional dyspnea in patients with COPD: evidence of the role of mechanical factors. Chest 2003;123:1794-1802.

[16] Iandelli I, Aliverti A, Kayser B, Dellacà R, Cala SJ, Duranti R, Kelly S, Scano G, Sliwinski P, Yan S, Macklem PT, Pedotti A. Determinants of exercise performance in normal men with externally imposed expiratory flow limitation. Journal of Applied Physiology 2002;92:1943-1952.

[17] Ward ME, Eidelman D, Stubbing DG, Bellemare F, Macklem PT. Respiratory sensation and pattern of respiratory muscle activation during diaphragm fatigue. Journal of Applied Physiology 1988;65:2181-2189.

[18] Potter WA, Olafsson S, Hyatt RE. Ventilatory mechanics and expiratory flow limitation during exercise in patients with obstructive lung disease. Journal of Clinicl Investigation 1971;50:910-919.

[19] Binazzi B, Bianchi R, Romagnoli I, Lanini B, Stendardi L, Gigliotti F, Scano G. Chest wall kinematics and Hoover's sign. Respiratory Physiology and Neurobiology. 2008;160:325-333.

[20] Garcia-Pachon, E. Paradoxical movement of the lateral rib margin (Hoover Sign) for detecting obstructive airway disease. Chest. 2002;122,651–655.

[21] Gilmartin JJ, Gibson GJ. Abnormalities of chest wall motion in patients with chronic airflow obstruction. Thorax. 1984;39,264–271.

[22]   Gilmartin, JJ, Gibson GJ. Mechanisms of paradoxical rib cage motion in patients with chronic pulmonary disease. American Review of Respiratory Diseases. 1986;134,683–687.

[23]   Chihara K, Kenyon CM, Macklem PT. Human rib cage distortability. Journal of Applied Physiology. 1996;81:437-447.

[24]   Jubran A, Tobin MJ. The effect of hyperinflation on rib cage-abdominal motion. American Review of Respiratory Diseases.1992;146:1378–1382.

[25]   Romagnoli I, Gigliotti F, Lanini B, Innocenti Bruni G, Coli C, Binazzi B, Stendardi L, Scano G. Chest wall kinematics and breathlessness during unsupported arm exercise in COPD patients. Respiratory Physiology and Neurobiology 2011;178:242-9.

[26]   Aliverti A, Quaranta M, Chakrabarti B, Albuquerque AL, Calverley PM. Paradoxical movement of the lower ribcage at rest and during exercise in COPD patients. European Respiratory Journal 2009;33:49-60.

[27]   Peters MM, Webb KA, O'Donnell DE.Combined physiological effects of bronchodilators and hyperoxia on exertional dyspnoea in normoxic COPD. Thorax 2006;61:559-67.

[28]   Vogiatzis I, Georgiadou O, Golemati S, Aliverti A, Kosmas E, Kastanakis E, Geladas N, Koutsoukou A, Nanas S, Zakynthinos S, Roussos C. Patterns of dynamic hyperinflation during exercise and recovery in patients with severe chronic obstructive pulmonary disease. Thorax 2005;60(9):723-9.

[29]   Tangri S, Wolf CR., The breathing pattern in chronic obstructive lung disease during the performance of some common daily activities. Chest 1973; 63:126-127.

[30]   Celli BR, Rassulo J, Make BJ. Dyssynchronous breathing during arm but not leg exercise in patients with chronic airflow obstruction. New England Journal of Medicine. 1986;314:1485-1490.

[31]   Ries AL, Bauldoff GS, Carlin BW, Casaburi R, Emery CF, Mahler DA, Make B, Rochester CL, Zuwallack R, Herrerias C. Pulmonary Rehabilitation: Joint ACCP/AACVPR Evidence-Based Clinical Practice Guidelines. Chest2007;131:4S-42S.

[32]   Janaudis-Ferreira T, Hill K, Goldstein RS, Robles-Ribeiro P, Beauchamp MK, Dolmage TE, Wadell K, Brooks D. Resistance arm training in patients with COPD: a randomized controlled trial. Chest 2011;139:151-158.

[33]   McKeough ZJ, Alison JA, Bayfield MS, Bye PT. Supported and unsupported arm exercise capacity following lung volume reduction surgery: a pilot study. Chronic Respiratory Disease 2005;2:59-65.

[34]   Holland AE, Hill CJ, Nehez E, Ntoumenopoulos G. Does unsupported upper limb exercise training improve symptoms and quality of life for patients with chronic obstructive pulmonary disease? J Cardiopulmonary Rehabilitation 2004;24:422-427.

[35] Ries AL, Ellis B, Hawkins RW. Upper extremity exercise training in chronic obstructive pulmonary disease. Chest 1988 ;93:688-692.

[36] Neiderman MS, Clemente PH, Fein AM, Feinsilver SH, Robinson DA, Ilowite JS, et al. Benefits of a pulmonary rehabilitation program: improvements are independent of lung function. Chest 1991;99:7989-7804.

[37] Martinez FJ, Vogel PD, Dupont DN, Stanopoulos I, Gray A, Beamis JF. Supported arm exercise vs unsupported arm exercise in the rehabilitation of patients with severe chronic airflow obstruction. Chest 1993;103(5):1397-402.

[38] Costi S, Crisafulli E, Antoni FD, Beneventi C, Fabbri LM, Clini EM. Effects of unsupported upper extremity exercise training in patients with COPD: a randomized clinical trial. Chest 2009;136:87-395.

[39] Couser JI Jr, Martinez FJ, Celli BR. Pulmonary rehabilitation that includes arm exercise reduces metabolic and ventilatory requirements for simple arm elevation. Chest 1993;103:37-41.

[40] Lake FR, Henderson K, Briffa T, Openshaw J, Musk AW. Upper-limb and lower-limb exercise training in patients with chronic airflow obstruction. Chest 1990;97:1077-1082.

[41] Romagnoli I, Scano G, Binazzi B, Coli C, Innocenti Bruni G, Stendardi L, Gigliotti F. Effect of arm training on unsupported arm exercise-related perception in COPD patients. Under revision in Respiratory Physiology and Neurobiology

[42] Bianchi R, Gigliotti F, Romagnoli I, Lanini B, Castellani C, Grazzini M, Scano G. Chest wall kinematics and breathlessness during pursed-lip breathing in patients with COPD. Chest 2004;125(2):459-65.

[43] Bianchi R, Gigliotti F, Romagnoli I, Lanini B, Castellani C, Binazzi B, Stendardi L, Grazzini M, Scano G. Patterns of chest wall kinematics during volitional pursed-lip breathing in COPD at rest. Respiratory Medicine 2007;101:1412–1418.

[44] Macklem PT. Exercise in COPD: damned if you do and damned if you don't. Thorax 2005;60:887-888.

[45] Nuss D, Kelly RE, Croitoru DP, Kats ME, A 10-year review of a minimally invasive technique for the correction of pectus excavatum. Journal of Pediatric Surgery 1998;33:545–552.

[46] Nuss D, Minimally invasive surgical repair of pectus excavatum. Seminars in Pediatric Surgery 2008;17:209–217.

[47] Binazzi B, Innocenti Bruni G, Coli C, Romagnoli I, Messineo A, Lo Piccolo R, Scano G, Gigliotti F. Chest wall kinematics in young subjects with pectus excavatum. Respiratory Physiology and Neurobiology 2012;180:211–217.

[48] Binazzi B, Innocenti Bruni G, Gigliotti F, Coli C, Romagnoli I, Messineo A, Lo Piccolo R, Scano G. Effects of the Nuss procedure on chest wall kinematics in adolescents

with pectus excavatum. In press in Respiratory Physiology and Neurobiology. DOI:
10.1016/j.resp.2012.05.015

Permissions

The contributors of this book come from diverse backgrounds, making this book a truly international effort. This book will bring forth new frontiers with its revolutionizing research information and detailed analysis of the nascent developments around the world.

We would like to thank Sergei L. Pyshkin, for lending his expertise to make the book truly unique. He has played a crucial role in the development of this book. Without his invaluable contribution this book wouldn't have been possible. He has made vital efforts to compile up to date information on the varied aspects of this subject to make this book a valuable addition to the collection of many professionals and students.

This book was conceptualized with the vision of imparting up-to-date information and advanced data in this field. To ensure the same, a matchless editorial board was set up. Every individual on the board went through rigorous rounds of assessment to prove their worth. After which they invested a large part of their time researching and compiling the most relevant data for our readers. Conferences and sessions were held from time to time between the editorial board and the contributing authors to present the data in the most comprehensible form. The editorial team has worked tirelessly to provide valuable and valid information to help people across the globe.

Every chapter published in this book has been scrutinized by our experts. Their significance has been extensively debated. The topics covered herein carry significant findings which will fuel the growth of the discipline. They may even be implemented as practical applications or may be referred to as a beginning point for another development. Chapters in this book were first published by InTech; hereby published with permission under the Creative Commons Attribution License or equivalent.

The editorial board has been involved in producing this book since its inception. They have spent rigorous hours researching and exploring the diverse topics which have resulted in the successful publishing of this book. They have passed on their knowledge of decades through this book. To expedite this challenging task, the publisher supported the team at every step. A small team of assistant editors was also appointed to further simplify the editing procedure and attain best results for the readers.

Our editorial team has been hand-picked from every corner of the world. Their multi-ethnicity adds dynamic inputs to the discussions which result in innovative outcomes. These outcomes are then further discussed with the researchers and contributors who give their valuable feedback and opinion regarding the same. The feedback is then collaborated with the researches and they are edited in a comprehensive manner to aid the understanding of the subject.

Apart from the editorial board, the designing team has also invested a significant amount of their time in understanding the subject and creating the most relevant covers. They scrutinized every image to scout for the most suitable representation of the subject and create an appropriate cover for the book.

The publishing team has been involved in this book since its early stages. They were actively engaged in every process, be it collecting the data, connecting with the contributors or procuring relevant information. The team has been an ardent support to the editorial, designing and production team. Their endless efforts to recruit the best for this project, has resulted in the accomplishment of this book. They are a veteran in the field of academics and their pool of knowledge is as vast as their experience in printing. Their expertise and guidance has proved useful at every step. Their uncompromising quality standards have made this book an exceptional effort. Their encouragement from time to time has been an inspiration for everyone.

The publisher and the editorial board hope that this book will prove to be a valuable piece of knowledge for researchers, students, practitioners and scholars across the globe.

List of Contributors

Florin Stanculescu
University of Bucharest, Bucharest-Magurele, Romania

Anca Stanculescu
National Institute of Materials Physics, Bucharest-Magurele, Romania

Vladimir G. Krasilenko
Vinnitsa Social Economy Institute of Open International University of Human Development "Ukraine", Ukraine

Aleksandr I. Nikolskyy and Alexander A. Lazarev
Vinnitsa National Technical University, Ukraine

Patrice Salzenstein
Centre National de la Recherche Scientifique (CNRS), Unité Mixte de Recherche (UMR), Franche Comté Electronique Mécanique Thermique Optique Sciences et Technologies (FEMTO-ST), France

Hiroki Kishikawa, Hirotaka Umegae, Nobuo Goto and Shin-ichiro Yanagiya
Department of Optical Science and Technology, The University of Tokushima, Japan

Yoshitomo Shiramizu and Jiro Oda
Department of Information and Computer Sciences, Toyohashi University of Technology, Japan

Ulrich H. P. Fischer
Harz University of Applied Sciences Friedrichstraße, Wernigerode

Moisés Rivas, Wendy Flores, Javier Rivera, Oleg Sergiyenko, Daniel Hernández-Balbuena and Alejandro Sánchez-Bueno
Engineering Institute of Autonomous University of Baja California (UABC), Mexico

Gui jiang Lin, Jingfeng Bi, Minghui Song, Jianqing Liu and Weiping Xiong
Xiamen San'an Optoelectronics Co., Ltd., China

Meichun Huang
Department of Physics, Xiamen University, China

Giulia Innocenti Bruni and Francesco Gigliotti
Section of Respiratory Rehabilitation, Fondazione Don C. Gnocchi, Florence, Italy

Giorgio Scano
Department of Internal Medicine, Section of Immunology and Respiratory Medicine, University of Florence, Florence, Italy
Section of Respiratory Rehabilitation, Fondazione Don C. Gnocchi, Florence, Italy

Printed in the USA
CPSIA information can be obtained
at www.ICGtesting.com
JSHW011400221024
72173JS00003B/358